DISSENT OVER DESCENT

DISSENT OVER DESCENT

Intelligent Design's Challenge to Darwinism

STEVE FULLER

Published in the UK in 2008 by
Icon Books Ltd, The Old Dairy, Brook Road,
Thriplow, Cambridge SG8 7RG
email: info@iconbooks.co.uk
www.iconbooks.co.uk

Sold in the UK, Europe, South Africa and Asia
by Faber & Faber Ltd, 3 Queen Square,
London WC1N 3AU or their agents

Distributed in the UK, Europe, South Africa and Asia
by TBS Ltd, TBS Distribution Centre, Colchester Road
Frating Green, Colchester CO7 7DW

This edition published in Australia in 2008
by Allen & Unwin Pty Ltd, PO Box 8500,
83 Alexander Street, Crows Nest, NSW 2065

Distributed in Canada by Penguin Books Canada,
90 Eglinton Avenue East, Suite 700,
Toronto, Ontario M4P 2YE

ISBN: 978-1840468-04-5

Typeset in 10.5pt Palatino by Marie Doherty

Printed and bound in the UK by
Clays of Bungay

Contents

Introduction

What is intelligent design (ID) theory? A snide but not inaccurate answer is that it is scientifically-credentialed creationism. ID theorists do little to hide their typically Christian religious inspiration, but they are steeped more in science than theology. Thus, they defend their position without reference to God or Scripture, but by the usual scientific appeals to reason and evidence.[1] As might be expected of a minority position with relatively little access to research funds, journal space and graduate students, ID theorists tend to reinterpret existing science rather than do original research. Their short-term goal is to justify room for alternative explanations for the emergence and maintenance of life on Earth to that of modern evolutionary theory, or 'genetically modified Darwinism'. However, ID's long-term goal is to reorganise the sciences so that biology and technology come to be treated as 'design sciences' in exactly the same sense, the former a science of God's design and the latter of human design. According to the ID theorist, technology imitates and – where possible – improves upon and perhaps even completes biology.

It is here that ID's creationism surfaces, specifically the idea that humanity is created 'in the image and likeness of God' – *in imago dei*, to recall St Augustine's 4th-century shorthand. Yet, contrary to the way in which its detractors depict it, ID is hardly a 'science-stopping' form of creationism. On the contrary, ID was behind the great Scientific Revolution that has been under way in the West since the 17th century, and it continues to provide the most powerful reason for doing science: nature is constructed so that we may understand and exercise dominion over it. Indeed, a central theme of this book is that it is much easier to justify evolution as the product of the history of science, than science as the product of evolutionary history. This is not to deny ID's own recent evolution from more explicitly biblical versions of creationism. In *Kitzmiller v. Dover Area School District* (2005), the only US court case in which ID tried (and failed) to justify itself as a part of the school science curriculum, a telling piece of evidence was that the proposed ID textbook was a warmed-over biblical creationist one. This proved, at least to the judge's satisfaction, ID's deceit in trying to pass off a religious doctrine – the teaching of which is not allowed in publicly funded schools – as a scientific theory.

The judge's rather po-faced verdict would have been greeted with perverse amusement by the founding fathers of the USA. They had a sufficiently arch sense of divine justice – or 'theodicy' – to recognise that 'private vices make for public virtue' through the alchemy of government. From the founders' Enlightenment standpoint, religion's propensity for dogmatic and fanatical attitudes is a 'private vice' that science, in lieu of an established state Church, then proceeds not to deny but to discipline, by forcing religion to replace articles of faith with logical and empirical demonstrations. Indeed, Francis Bacon had invented the scientific method 150 years earlier, precisely to sublimate the potential for violence among religiously inspired inquirers. However, Bacon had no illusions that scientists' religious motives would entirely disappear. Their lingering differences, albeit increasingly refined over time, ensured that inquiry would remain dynamic: each would always keep the others' epistemic ambitions in check.

It is always tempting to read intellectual history from front to back, rather than vice versa. We tend to think of the target of time's arrow as wherever we happen to be in history, with much of the intervening period marked by diversion and delay. On this view, something like the Neo-Darwinian account of evolution has always been our collective intellectual destination – after all, is it not the truth? But its time of arrival has been needlessly delayed by various creationist roadblocks, the latest being ID. However, this is not the most productive way for the past to inform the future. One of the few things on which agreement can be reached among the scholars working in my own field – the interdisciplinary study of the history, philosophy and sociology of science known as 'science and technology studies' (STS) – is what the Harvard philosopher Hilary Putnam originally called a 'pessimistic meta-induction'; to wit, the bigger the theory, the harder it falls. Putnam meant that revolutionary change in science tends to occur at the level of general explanatory frameworks: grand theories eventually overreach their grasp. It happened to Newton (courtesy of Einstein), and it is very likely to happen to Darwin. After the fall, the facts subsumed by these grand theories remain largely intact, but they come to be interpreted in a systematically different way.

In contrast, to a properly 'back to front' view of intellectual history, the best strategic reason to stick with the scientific orthodoxy runs no deeper than insurance against risk. Imagine you are what Thomas Kuhn called a 'normal scientist', someone trained to solve puzzles posed by the dominant, if not only, respectable research programme, or a 'paradigm'. If the paradigm persists over the course of your career, you will probably end up having made some modest positive contributions. But suppose a new paradigm comes to the fore: your inevitable obsolescence will be excused for having been a product of its times. Now, step back from this logic. Who says risk aversion is the most rational strategy for intellectual life, especially for professional academics who still enjoy the luxury of testing the frontiers of knowledge without their livelihoods hanging on what follows? I allude here to lifelong tenure, which – insofar as it still exists – is normally justified in terms of allowing one the freedom to work on socially irrelevant problems that fellow professionals happen to value. However, this undersells tenure's historic role in emboldening academics to challenge nostrums that cut across their own field and society at large.

The book before you begins by challenging the taken-for-granted idea that there is a consensus of opinion in the scientific community. This alleged fact alone was sufficient to disqualify ID as science in *Kitzmiller*. Nevertheless, the very idea of a scientific consensus is not only antithetical to the spirit of free inquiry, but as a matter of fact it may not even exist. In chapter 2, I deal with the emergence of science as an intellectual project and the scientist as a social role. Here I highlight the centrality of the idea of humans as creatures *in imago dei* for whom nature is therefore 'intelligible'. This belief, common to the Scientific Revolution and the Enlightenment, had been originally suggested by a strong reading of St Augustine's views of biblical interpretation. However, the evidence of pointless death in nature weaned Darwin away from these views, which in the end led him to reject the title of 'scientist'. In chapter 3, I consider the origins and legitimacy of the science–religion conflict, focusing on various peacekeeping efforts that travel under the name of 'theistic evolution'. I treat these with considerable scepticism, except for the version advanced by Pierre Teilhard de Chardin, whose subtle Neo-Lamarckian version of 'creative evolution' engaged the founders of the amalgamation of Darwin's evolutionary account of natural history and lab-based population genetics, including its basis in molecular biology, that is now routinely called the 'Neo-Darwinian synthesis'. I believe Teilhard's views are worth revisiting today, in light of developments in biology since his death more than half a century ago.

Chapter 4 looks at the various attempts to frame the difference between evolution and ID in terms of the science/non-science distinction. All of these efforts fail mainly because ID's alleged non-scientific activities trade on fundamental unclarities in the scientific grounding of evolution itself. In this context, the structure and history of astrology and evolution as grand theories provides an interesting point of comparison. Chapters 5 and 6 are concerned with the history and persistence of design-based thinking in biology, from the 17th-century theological science of theodicy to its secular descendants in the various projects associated with 'biophysics' in the 20th century, not least information theory and molecular biology, the latter now the centrepiece of the Neo-Darwinian synthesis. All of these presume various ideas of optimisation and cosmic engineering that in many cases retain elements of their biblical inspiration. Chapter 7 deals straightforwardly with how modern scientific developments associated with ID can be seen as having arisen from a certain understanding of biblical literalism. In the conclusion, I propose a general strategy for ID to improve its position in the current debate with evolution. My partisanship comes from my service as rebuttal witness for the defence in *Kitzmiller*, more of which is discussed in a recent book of mine, *Science vs. Religion? Intelligent Design and the Problem of Evolution* (Polity Press, 2007).

The argument of this book, and my style more generally, is punctuated – some might say interrupted – by much jumping back and forth across the centuries, treating the past as contemporaneous with the present. The overall effect may appear wilfully diversionary. I plead guilty. My point is to reveal the spatio-temporal seams hidden in the invariably 'syncretistic' character of intellectual discourse. 'Syncretism' is a term from linguistics for grammatical forms that result from combining historically distinct roots, which to the unsuspecting mind leaves the impression of a timeless idea. A much noted 19th-century example was Auguste Comte's coinage of 'sociology', whose mix of Latin and Greek roots immediately threw into doubt the legitimacy of the new field, given the patent differences between the cultures whose words had been combined. But syncretism extends beyond the constitution of words to our very conception of science. The most influential general theory of science of at least the last half-century, Kuhn's account of paradigms, is based on a pastiche of images of scientific practice drawn from the early 17th to the early 20th centuries that happened to resonate with the self-understanding of late-20th-century scientists and, more importantly, those who justified science's exalted position as a form of knowledge in society.[2]

Science conceals its syncretism by a 'winner-takes-all' attitude to knowledge production and the 'front to back' approach to history that this breeds. This is an insight for which Kuhn is rightly credited. Someone wishing to advance the dominant scientific paradigm is presumed entitled to appropriate the work of any scientific predecessors, regardless of the projects that drove their scientific work. Thus, a staunch atheist like Richard Dawkins can, in good conscience, draw on such religiously inspired figures as Carolus Linnaeus, Gregor Mendel, Ronald Fisher, Sewall Wright and Theodosius Dobzhansky, because their scientific contributions have been absorbed by the Neo-Darwinian synthesis without incorporating their specific religious interests. The ongoing resistance to ID shows that the same courtesy is not extended to non-paradigmatic research agendas, even when they might breathe new life into those original, religiously motivated projects.

The point of my open syncretism, then, is to force the reader to re-examine the original terms in which various insights have been incorporated – or not – into the scientific canon. Before the pursuit of science was so bound up with material resources and modes of legitimation, it was natural to presume that scientific theories, concepts and findings may be made and used by any intellectual project, unless specifically prohibited. Unfortunately the burden of proof has now been reversed, as the orthodoxy in science – misleadingly called the 'scientific consensus' – requires not only competence but also loyalty before licensing access to its knowledge base. This results in a false but popular image of evolution as the natural culmination of the history of science, and ID as the most recent and perhaps cleverest incarnation of the counter-tradition that has always threatened science. As a matter of fact, the dominant trajectory promoting science in the West has been strongly grounded in its monotheistic traditions, of which ID is a natural successor. We shall see in these pages that evolution, including its greatest theorist Charles Darwin, has had a difficult relationship with these traditions.

Before launching into the argument, a few personal remarks are in order. I want to thank Simon Flynn for the great faith and patience he has displayed in this project. Howard Cattermole, Andrew Hegarty and Peter Loose have provided me with hospitable forums to develop these ideas more fully. To those not familiar with my previous work, I should say that I am a secular humanist who has been steeped in the historical and philosophical relations between science and religion since my school days with the Jesuits, the subtle masters of reconciling the seemingly irreconcilable. To this day I regard the difference between science and religion as more institutional than intellectual: they are basically trading on the same ideas but pursuing them by different means and judging them by different standards. While I cannot honestly say that I believe in a divine personal creator, no plausible alternative has yet been offered to justify the pursuit of science as a search for the ultimate systematic understanding of reality.

Even if most scientists nowadays call themselves atheists, atheism as a positive doctrine has done precious little for science. Those scientists who happily trade on their atheism may justify science in one of three equally inadequate ways: first, they might point to science's practical benefits, both intended and unintended. But as the two world wars in the last century made clear, this justification makes science a hostage to fortune, resulting in periodic antiscientific backlashes. Second, scientists may appeal to subjective aesthetic factors as motivating their craft. While that may suffice for the scientific practitioner, it does little to justify the increasing cost (both intended and unintended) to the society supporting his or her activities. Finally, like Richard Dawkins, scientists may simply identify 'atheism' with a secular version of the theological justification of science. It is the last justification to which the reader should be especially attentive in the following pages.

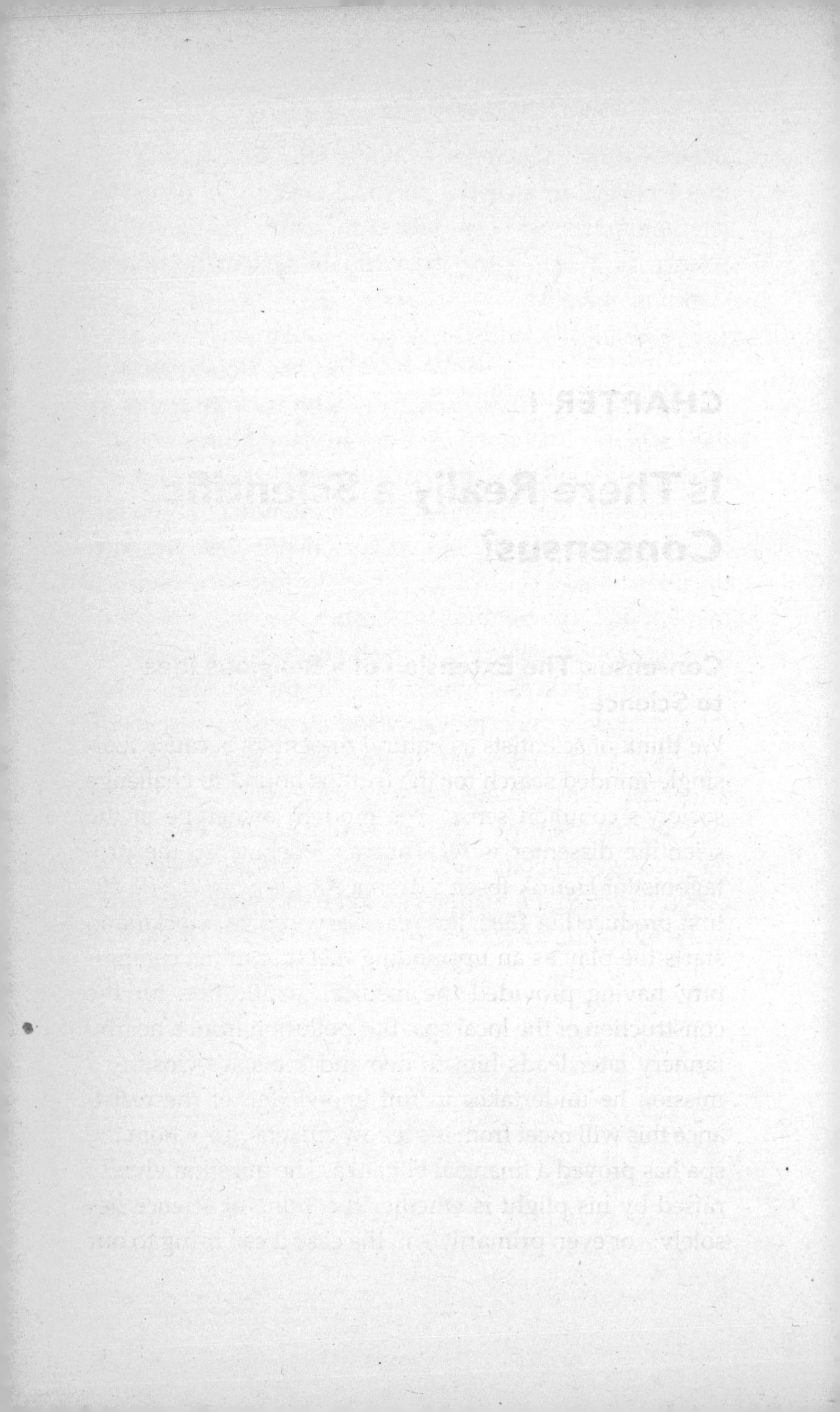

CHAPTER 1

Is There Really a Scientific Consensus?

Consensus: The Extension of a Religious Idea to Science

We think of scientists as natural dissenters because their single-minded search for the truth is bound to challenge society's common sense. The modern archetype of the scientific dissenter is Dr Thomas Stockmann, the protagonist of Henrik Ibsen's drama *An Enemy of the People*, first produced in 1882, the year Darwin died. Stockmann starts the play as an upstanding member of his community, having provided the medical justification for the construction of the local spa. But pollution from a nearby tannery later leads him to demand the spa's closure, a mission he undertakes in full knowledge of the resistance this will meet from his fellow citizens, for whom the spa has proved a financial bonanza. The question vividly raised by his plight is whether the value of science lies solely – or even primarily – in the ease it can bring to our lives, or in its capacity to challenge the views we take for granted. Stockmann's real-life descendants were easy to identify by the mid-20th century – for example, the US ecologist Barry Commoner, who mobilised against the proliferation of nuclear arms, the militarisation of science and the despoliation of the physical environment.

However, the practice of dissent is neither exclusive nor original to science. It is ultimately grounded in religion – specifically, those religions descending from the biblical Abraham: Judaism, Christianity and Islam. Abraham decided to follow what he took to be the voice of The One True God instructing him to disown his society, leave his land and sacrifice his son (Genesis 22). From a secular standpoint, the element of blind faith stands out in Abraham's decision. But the theologically salient point is that Abraham refused the path of least resistance, which would have had him conform to local traditions to continue enjoying his secular success. Instead he opened himself to a broader existential horizon, on the basis of which he decided to risk alienation from everything familiar to him. But of course Abraham's decision was not made for everyone for all time: it must also be taken by each person who would follow in Abraham's footsteps.

Dissent is not merely possible but openly invited once belief in The One True God is treated as a matter of personal decision that may fail to secure the acceptance of fellow believers. Thus, Christian dissenters differ from, say, sceptics and atheists in regarding themselves as Christian to at least the same extent as those who would persecute and exile them. However, dissenters typically practise their Christianity differently or draw different conclusions from Scripture, all of which threatens a religion whose strength comes from a common orientation to God. Indeed, communion with God is the historic source of communal solidarity more generally, a foundational insight in the discipline of sociology. Little surprise, then, that God's great antagonist, Satan, is often portrayed in literature as a being whose power increases from the confusion and strife that dissent invariably brings. Those who would sow seeds of dissent in the faith community have therefore been treated as messengers of Satan.

The desire for communion, now in the guise of consensus, applies to science today, with 'Truth' having become secular shorthand for 'The One True God'. Science is no longer a source of religious dissent – as it arguably was in the cases of Galileo or Darwin. Rather, dissent has beset science itself. Dissent *within* science is so unusual because it presupposes a *scientific consensus*, an orthodoxy around which the scientific ranks may close.[1] Yet, for much of its history, science has distanced itself from religion precisely by its refusal to adopt an institutional structure – an established Church – that would sharply distinguish between permitted and prohibited beliefs.[2] Karl Popper translated this historical observation into a philosophical point about the difference between 'closed' and 'open' societies, the latter exemplified by science. Philosophers of science still distinguish between contexts of 'discovery' and 'justification', whereby the scientist is permitted any starting point (the context of discovery) as long as it results in conclusions that withstand serious logical and empirical examination (the context of justification). In practice, this has meant that scientists may privately think whatever they wish, but their public knowledge claims must sustain the scrutiny of scientists with quite different beliefs and motives. Francis Bacon invented the scientific method amid the wars of religion in early 17th-century Europe, to enable inquirers operating with radically different world views to settle their differences in a fashion that would allow them to benefit from a body of empirical knowledge that could command the respect of them all.[3]

The sensibility just described contrasts strikingly with a common religious attitude, defended by philosophers as different as Blaise Pascal, William James and Ludwig Wittgenstein – namely that to have even a chance of finding God, one must first lead a godly life. In secular translation: there is a consequential connection between already believing certain things – and hence behaving in certain ways – and one's receptiveness to the truth, should it ever be revealed. The ultimate source of this thesis is St Augustine, whose exhortation *crede ut intellegas* ('Believe in order to understand') struck a balance between his philosophical scruples, which could not guarantee knowledge of the truth by any means, and his episcopal office, through which he aimed to consolidate the Christian community in the tempestuous heresy-filled days of the early Church. In all matters of the Christian faith, St Augustine remains – even after over 1,500 years – an interesting place to begin. In chapters 2 and 7, we shall return to his views on biblical literalism.

Science began to adopt the Augustinian line only in the 20th century. This was mainly through proactive invocations of Louis Pasteur's maxim 'Discovery favours the prepared mind.' 'Preparation' came to be seen as more than happening to be in the right place at the right time: it became an opportunity that is afforded only once a certain form of discipline is undertaken. Highly influential philosophies of science were subsequently built upon this observation, notably Michael Polanyi's and especially Thomas Kuhn's.[4] However, to a dissenter, a mainstream scientist's 'paradigm' looks like tunnel vision, if not groupthink. The classic mechanism for the manufacture and reproduction of any intellectual orthodoxy has been mandatory common training for a substantial period. This is the sense of 'discipline' that joins the medieval monastic orders with today's professional scientists – but *not*, say, the wealthy amateur Charles Darwin, the patent officer Albert Einstein and the host of cross-disciplinary migrants and 'unprofessional' practitioners responsible for most innovative research in science prior to the last couple of generations. Indeed, the varied social and intellectual backgrounds of innovative scientists, at least until the mid-20th century, refute the claim that a strong sense of common mental preparation has been the historic norm, though as we shall see in the next chapter the need for such preparation inspired the invention of 'scientist' as a dedicated profession.

The very idea of a scientific consensus, to which one might point as the 'pole star' of the intellectual world, was socially constructed during the Cold War. Previously the collective mind of science could be characterised as drawn in certain directions, but typically these were multiple and reversible. Before the Cold War, power in science had never been so centralised as to inhibit or deform the expression of dissent. Of course, national systems of education, professional academic societies and peer-reviewed journals were already in place. However, they were not tightly bound together through funding mechanisms. National education ministries might foster a common disciplinary identity through the adoption of approved textbooks but their control, except in the case of imperial powers, rarely extended beyond national boundaries. Thus, at the dawn of the First World War, the historian and philosopher of science Pierre Duhem could still make epistemologically interesting remarks about the national styles of French, German and English 'science', meaning *physical* science. While we might still observe national differences today in a discipline like literary studies or sociology, there would be little point in repeating Duhem's exercise for physics or biology – or, for that matter, psychology or economics.

The modern period has witnessed many attempts to forge a worldwide scientific consensus of opinion, most of these of a self-protective character. Scientists, however employed, routinely think of themselves as producing knowledge of potentially universal import. But this then raises the question of how the necessarily far-flung nature of such work is to be protected from various forms of local interference, ranging from relatively passive obstacles like language differences to more active ones like economic rivalries and political hostilities. From the 17th century to the present day, the answers have varied. They have included proposals for global scientific languages, international treaties of free scientific exchange and Marx-inspired movements that would instil a common class consciousness in the world's scientific workers. All were promising for a limited period, during which they often did some good and prevented some harm. But they all failed as mechanisms for consolidating opinion on substantive scientific matters.

Germany is usually seen as having set the pace for the centralisation of scientific authority, starting with unification in 1870. Perhaps the most enduring of its institutional innovations has been the 'partnership' of industry and academia under the watchful eye of the state in the Kaiser Wilhelm Institutes, precursors of today's Max Planck Institutes. But despite its preoccupation with knowledge production as an instrument of global dominance, Germany never delegated the task of defining the research frontier to supposedly self-organising academic disciplines. If anything the German tendency was, for better or worse, to undermine the power of the professional scientific bodies that nowadays promote the idea of a scientific consensus. Thus, Bismarck's visionary higher education minister, Friedrich Althoff, catapulted the Second Reich to the top spot of scientific nations in a single generation by refusing to allow pleas for disciplinary continuity to mask attempts by professors to anoint their successors. Althoff held that academic autonomy is best preserved by periodically shaking up the local intellectual environment, thereby forcing academics to redefine themselves in new ways. Half a century later, Hitler secured the support of most non-Jewish academics in the Third Reich by permitting them to follow the spirit of empirical inquiry wherever it led, which in the case of the social and biomedical sciences resulted in experiments that would have been prohibited in a regime that protected human subjects via discipline-based codes of professional conduct. Those who remained active researchers in Nazi Germany understandably thought that their sphere of freedom was wider, not narrower, than it had been under the democratic but 'politically correct' Weimar Republic.[5]

The groupthink that emerged among post-war scientists radically differed from the social psychology of Nazi scientists who seemed to see themselves as free researchers. Most successful scientists on both the capitalist and communist sides of the ensuing Cold War divide fully realised that they served two masters at once, their discipline and their state. This reflected two distinctive developments in science in the second half of the 20th century: increased technical specialisation and long-term agenda-setting. The former has enabled research to enjoy a dual existence as peer-review publications for academic colleagues and as confidential reports for defence analysts and policymakers. The latter reflected the Cold War's unique preoccupation with anticipating and pre-empting the moves of opponents, which in turn played to the scientific propensity for speculative thinking, as epitomised in scenarios and simulations. The result was that both scientists and politicians found it easy to merge the physics-driven quest for the mathematical unity of natural knowledge with the military interest in harnessing the inner workings of the atom and the farthest reaches of outer space. The universal validity and relevance of science came to be tied inextricably to the prospect of mutually assured destruction, as both were predicated on a knowledge of the fundamental forces of nature. In this way, the mushroom cloud of atom-based weaponry emerged, both literally and figuratively, out of the blue skies of atom-based research.

After Two World Wars: Two Science Wars

Today Cold War science policy evokes fear and loathing, but future historians may liken it to the 'Pax Britannica' that governed international relations before the onset of the First World War. While it has been long assumed that the history of science should be told as a succession of theories, paradigms, research traditions and disciplines, it was only during the Cold War that science came to be formally organised on a global scale to reflect this narrative logic. The benchmark was set by the US National Science Foundation (NSF), established in 1950 with discipline-based directorates that extended the scope of peer review beyond the authorisation of publication to the very licensing of research through a generous grant scheme, which after Sputnik was extended to cover undergraduate and postgraduate training. Never before had the difference between orthodox and heterodox scientific practice been so clearly institutionalised. Indeed, sociologists of science borrowed a contrast from international development studies – 'core' versus 'periphery' – to underscore the invidiousness of this institutionalised consensus. This was the context in which Kuhn developed his influential view of science as a highly centralised, even authoritarian, paradigm-driven enterprise.

Moreover, an early NSF seed grant went to the chemist Eugene Garfield for the prototype of the Science Citation Index (SCI), a computerised cross-referenced database of all scientific journal articles. It was justified on national security grounds – specifically, to get the edge on the USSR's rumoured avoidance of duplicated effort through a centralised state planning board for science policy.[6] In line with socialist thinkers of the time, Garfield supposed that a diligent state apparatus could remove the barriers to knowledge flow that would otherwise develop in a market-based economy, where competitors could gain advantage over each other simply by withholding information. Regardless of the ultimate truth of this perspective, certainly by the 1960s the historian Derek de Solla Price had begun to persuade a large Western audience that citation counts are fairly treated as a running account of science's collective consciousness.[7] This empowered policymakers to conceptualise science as an organism whose vital signs could be monitored and ministered through the SCI. To his credit, Price made it clear that the SCI provided evidence only for aggregate statistical tendencies, *not* individual performance. But these instruments of Cold War science policy were eventually turned on the scientific community that initially supported them, as evidenced by the increasing reliance on these 'metrics' in the periodic Research Assessment Exercises to which academic staff in the United Kingdom and elsewhere have had to submit.

The end of the Cold War finally forced what had become massive 'welfare–warfare states' on both sides to face up to mounting budget deficits. National security no longer automatically justified increased defence spending, even for scientific research. The defining moment of this sea change occurred in 1992, when the United States Congress cancelled the budget for the Superconducting Supercollider (SSC) project, which would have built the world's largest particle accelerator underneath Texas. This event may be seen as formally marking the onset of the first of the modern science wars, or 'Science War I'.

At the time, scientific proponents of the SSC, especially the Nobel laureate Steven Weinberg, blamed the corrosive influence of science and technology studies (STS) on policymakers.[8] STS is an interdisciplinary field that largely arose from some trained scientists (including engineers) who, in the light of science's dubious political and economic entanglements, disconnected themselves from the official story of science's exemplary objectivity. By 1992, STS had been engaged in almost twenty years of debate with philosophers over whether the historical and sociological record of scientific practice lived up to its distinctive epistemological reputation for objectivity. The aspect of this debate that seeped into science policy circles was a general scepticism about science as a unified mode of inquiry with a uniquely progressive character.

As it turned out, the research wing of Congress, the Office of Technology Assessment (OTA), was staffed by people sympathetic to the STS perspective. They refused to take the testimony of Weinberg and other prominent physicists at face value as being representative of the entire physics community. Instead they canvassed a wide range of indicators about the state of physics, discovering along the way that the proposed SSC would benefit only a small number of elite practitioners in highly speculative ventures. That was sufficient for Congress to kill the project in good conscience, a feat of 'cognitive euthanasia'.[9] Senior scientists retaliated with a steady stream of books and articles questioning the competence of historians, philosophers and sociologists to evaluate the state of science. This process culminated with the so-called Sokal hoax, whereby a physicist managed to get an article, laced with nonsensical scientific claims but cloaked in politically correct jargon, published in a leading cultural studies journal. The hoax made the front page of the *New York Times* on 18 May 1996, resulting in at least five years of embarrassment for the STS community.[10] Generally speaking, STS responded by closing ranks and adopting the rhetoric of professionalism, emphasising its own distinctive expertise as separate from that normally gained by ordinary scientific training. In the short term, the scientists seemed to have got the upper hand over their critics, broadly portrayed as 'academic leftists' who would subvert all standards of truth.[11] However, the *secularisation of science* had already set in.

The word 'secularisation' is used partly to capture the role that science played in demystifying religious authority from the 17th to the 19th centuries. By the dawn of the 21st century, however, science had clearly shifted in status from an *agent* to a *target* of secularisation.[12] The balance of power for determining the course of science had shifted from professional to lay groups, just as the spread of literacy and publishing in the 16th century enabled Christians to wrench control of their religion from the Catholic clergy. Similarly, in our own day, the expansion of the internet as a source of alternative expertise has emboldened people to adopt a 'pick and mix' consumerist attitude toward science. This has fostered a renascence of so-called 'New Age sciences', like homeopathy, that eclectically combine up-to-date scientific knowledge and traditional practitioner wisdom. But it applies equally to the spread of ID. The chair of the Dover, Pennsylvania school board responsible for introducing the theory into the high school science curriculum claimed that she was persuaded of its credibility as an alternative to evolution after surfing the internet.[13]

Just as the Protestant denominations differed over the Catholic doctrines they accepted, rejected and reformed, so too the public's interest in science today is both increasing and increasingly discriminating. This explains two countervailing tendencies: an all-time *high* in popular science readership and an all-time *low* in academic scientific enrolments. Despite the pervasive fretting over the latter, the fact remains that science – both its ideas and its applications – is increasingly used to define beliefs, attitudes and stakes in larger societal issues. However, surveys of scientific attitudes specifically and social attitudes more generally fail to reflect this comprehensiveness of influence. In particular, surveys on specifically scientific attitudes tend to be tailored to suit client interests, be they those of the Royal Society or Monsanto. Consequently, we really have no clear sense of the extent to which so-called expert and lay opinion differ on the scientific issues that increasingly frame public debates. And if there are differences, where do they emerge, especially given the widespread view that experts know more and more about less and less? Given the opportunity to address these questions, it is quite likely that elite and rank-and-file scientific – let alone scientific and non-scientific – judgement might diverge, including over evolution versus ID. In any case, it is much too convenient to blame the mass media for misrepresenting an aspect of public life that has been always subject to a free-for-all of rival spin doctors.

In the past two decades, Science War I has yielded to Science War II. If Science War I was focused almost exclusively on the post-Cold War meltdown in national scientific research agendas, Science War II reflects an additional devolution in the sources of public information, including education, about science. In this context, the United States' constitutional delegation of responsibility for the primary and secondary school curricula to local taxpaying units, subject to state oversight, has become the latest field of engagement, the most publicised set of cases involving the introduction of ID alongside Darwin's theory of evolution in biology classes. Indicative of the transition from Science War I to Science War II is the overall disciplinary shift in science's centre of gravity from physics to biology, as measured in terms of research funding, student interest and media coverage. It has even influenced philosophical and sociological conceptions of what it is to be a science. If physics unifies science, biology diversifies it.[14] In particular, the range of specialities subsumed under contemporary biology – from palaeontology through field ecology to population genetics, molecular biology and most recently nano-level biotechnology – do not appeal equally to all scientific, political and economic interests.[15]

This diversity is reflected not only in terms of how the science of biology is organised but also in the nature of the entities it studies. Indeed, Darwin's clearest mark on biology is the field's preoccupation with what differentiates, rather than unifies, organisms, including humans. Thus, racism received a new lease of life once biologists adopted a narrative centring on the fitness of species to survive in environments largely (though not entirely) outside their control. In this context, races were first redefined as proto-species but then, as knowledge of genetics grew, the traditional category of race was itself deemed to be an unreliable indicator of substantive genetic differences. Such differences, now often travelling under the politically correct banner of 'genetic diversity', continue to matter as much as ever – if not more so. However, the spin given to what may come to be seen as 'micro-racism' is radically different: the ideal of racially homogeneous nation-states closed under a common welfare system (as realised in Scandinavia and to which Germany aspired in the 20th century) has been replaced by an ecology of niche markets through which a variety of biomedical products and treatments target a genetically diverse population.[16]

The Docility of Scientists and the Opacity of Their Opinions

In the face of science's increasing heterogeneity, the continuing elitism of the scientific establishment is striking. Consider the UK Royal Society (RS) and the US National Academy of Sciences (NAS), which are agreed that evolution is science and ID is not. Membership of these august bodies is by invitation only. They may present themselves, and are usually taken, as speaking for the scientific communities of their respective nations. But these bodies are not 'representative' of the totality of scientists in either how they are constituted or to whom they are ultimately accountable. The Royal Society's charter is defined primarily in terms of mutual non-interference between itself and the state, while the National Academy of Sciences is specifically commissioned to do research for, and offer advice to, the US government. Moreover, neither the RS nor the NAS has ever shown any systematic interest in what rank-and-file scientists think about scientific matters. Of course, they have periodically conducted surveys on their own members. Indeed, one recent survey revealed that 85 per cent of NAS members are atheists, despite living in one of the world's most openly religious nations. It speaks volumes that the pressing question arising from the report was not why the NAS appears to be unrepresentative of the US population but why the remaining 15 per cent of its members are not atheists as well.[17]

What is most striking in all this is the docility of rank-and-file scientists in the face of their lack of formal representation. In this respect, the politics of Orwell's *Animal Farm* applies to science: in the 'peer review' culture of science, some scientists are more entitled to peerage than others. The closest political analogue to this situation may be the ancient idea of representation proportional to size of property holding, where a scientist's citation count (i.e. the number of times other scientists cite that scientist's work) functions as the size of her holding, or degree of her possession, of a research field. This in turn tends to be positively correlated with institutional reputation.[18] To be sure, the 'one man, one vote' principle is now the politically correct version of democratic decision-making. Yet, the ancient idea lingers not only in the organisation of science but also an idea, still popular in some business quarters, that would turn over the management of the economy to the most successful producers because they have both the best track record and the greatest stake in the economy doing well. In its most benign form, this 'corporatist' approach to state management was a mainstay of the welfare states of Germany and Scandinavia. Less benign versions evolved into fascism in the first half of the 20th century.

Aside from science, it is difficult to imagine another sector of society in our time – the Roman Catholic Church perhaps – in which such elite corporatist arrangements have been tolerated as grounds for popular authority. What would happen if rank-and-file scientists organised themselves to oppose a stand taken by an elite scientific body? Would such a body give formal recognition to a petition, let alone a referendum? As with the Church, a large element of mystification is at play: scientists in ordinary careers overestimate the cognitive differences between themselves and their status superiors, resulting in an amplified sense of the latter's span of competence. To be sure, it is a thankless task to draw attention to the fallibility of such celebrated scientists as Einstein, Bohr, Heisenberg or, in our own day, Steven Weinberg, once they stray beyond their specialities. (The task has been assumed – perhaps a bit too enthusiastically – by those murdered messengers, STS scholars.) Not surprisingly, then, less distinguished scientists seem inclined to presume that their elite colleagues are entitled – perhaps even required, as a kind of *noblesse oblige* – to leverage their expertise in matters relating to politics, religion and the overall trajectory of science itself.

Nevertheless, successful scientists are susceptible to exactly the same cognitive liabilities that afflict successful people in other fields. They tend to generalise superstitiously about the causes of their success, underestimating the element of chance involved and the differences between their own field and fields that they would have learn from their example. Of course, none of this detracts from their proven accomplishments. But whether these scientists are mentally equipped to plot the future course of science more generally is an open question. Indeed, there may be a psychic trade-off between contributing to short- and long-term policy thinking about science. The short term may be best served by the advice of the most successful scientists, who are well placed to exploit the full potential of the research trajectories to which they have personally contributed. But the long term may be better served by those who, regardless of their personal scientific accomplishments, understand science from a broader temporal horizon. After all, even the most fruitful research programme began life as a decision from among alternatives, versions of which may be worth resurrecting now, if the path originally taken is delivering diminishing returns on investment or simply tying up too many resources that could be used for other worthy projects.

In the last few years, this potential conflict between science policy horizons has become more visible, as the RS and NAS have closed ranks in the face of increasingly vocal opposition to the orthodoxies that have hardened around climate change and evolutionary theory. A bellwether is the recent retranslation of the Royal Society's Latin motto, *Nullius in verba*, traditionally rendered as 'On the word of no one', an exhortation not to trust authority but to examine matters for oneself. When the relevant authority was the Church, whose *modus operandi* was the interpretation of sacred texts, this translation served as a convenient way of limiting the reach of religion into the domain of science. But nowadays that seems not to be enough. In 2007, Robert May – distinguished mathematical ecologist, former RS president and scientific advisor to both Conservative and Labour governments – entitled a lead article in the *Times Literary Supplement* with his own updated Orwellian gloss: 'Respect the facts.'[19] Ironically, only five years earlier RS fellows had bristled when Tony Blair tried to persuade them that science served society best when it provided 'facts' that could feed into an 'evidence-based' policy process.[20] At that time, the RS stressed the open-endedness of scientific inquiry. While clearly things have changed, one constant has been the failure of elite scientific bodies like the RS and NAS to consult rank-and-file scientists, let alone in a systematic fashion that might result in an analogue to the regularly-administered British Social Attitudes Survey.

What's ultimately missing here is obvious: any of the rather different senses of democratic transparency provided by votes, prices and ratings. These are outcomes of processes – elections, exchanges and airings, respectively – whose fairness rests on both producers and consumers being allowed to exercise independent judgement in what is sold and bought. Where this condition fails to be met, the concentration of power is easily mistaken for a consensus of opinion. Nevertheless, it is easy to imagine some of the areas in which one might wish periodically to sample scientists and non-scientists about scientific matters, asking them to relate general attitudes to current issues, including the following:

Subjects would be asked identifying questions about

- geographical location (list of regions given);
- field of expertise (defined as desired; by education, practice etc.);
- context of practice (education, research, policymaking etc.);
- if researcher, how they work (laboratory, field, computer modelling, library-based) and how they are employed (in academia, industry, government, NGO, self, other);

- if policymaker, at what level they work (supranational, national, state, local);
- if educator, what level (primary or secondary school, undergraduate, graduate);
- main source of information about various current science-based issues (scientific publications, policy reports, popular media);
- closeness to what they are commenting on (researcher, decision-maker, advisor, spectator).

Subjects would also be asked substantive questions

- about how their own views differ from what they believe most scientists and non-scientists think about a variety of general and specific scientific topics, e.g. the scientific method, multiple universes, evolution, global warming, genetically modified foods;
- about how much influence politics, business and religion do and should have over, on the one hand, the scientific research agenda and, on the other, science education;
- about if and when lay people (taken both individually and collectively) should defer to expert scientific judgement;
- about if and how they distinguish 'competence' from 'expertise' in public discussions of various science-related matters, e.g. under what circumstances people not formally trained in a given area should be able to comment responsibly in the media about it;
- about how they understand the 'internal' and 'external' politics of science, e.g. whether they are or ought to be 'democratic';

- about the larger social prospects for various novel technologies, both in terms of their realisation and their likely costs and benefits;
- about the need for 'limits' on scientific research, application, dissemination etc.

Although there is no reliable empirical basis for making claims about a 'scientific consensus', it is easy to identify who exercises power in science. The RS and NAS commission research on the basis of which they deliver *ex cathedra* judgements, which are subject to approval by a body no larger or more representative than their own governing boards. In effect, they are oligarchies that exercise power without ever having to impose force, the envy of any self-respecting mafia. The rational response of professional scientists not privy to this process is to consent to a *second-order* judgement that the RS and NAS are the right places to turn for a *first* opinion about a given scientific matter, in the hope that more than one informed opinion is to be had. Thus, the onset of the science wars has generated issues-based opposition groups, including some of international scope. However, these groups – typically NGOs oriented towards ordinary citizens or the environment and associated with the political left – have been careful not to undermine the appearance of a scientific consensus. After all, their own moral authority capitalises on the failure of politicians to follow through on a putatively united scientific front. As for those inclined to oppose the very idea of a consensus of scientific opinion, they typically hail from right-leaning think tanks whose dissent fails to remove the appearance of self-serving moves by marginal scientific players. Thus, organised science nowadays is without a loyal opposition entrusted with the task of regularly reminding both scientists and the public that the sheer concentration of power is not a legitimate mode of securing consent.

The Social Construction of Evolution's Scientific Consensus

In terms of the science-based issues raised in these pages, the above survey should be able to distinguish whether or not a scientist believes in evolution, and whether or not he or she believes that a belief in evolution is cognitively necessary in order to do creditable biological research. The distinction is important because, as a matter of fact, the vast majority of published research in the biomedical sciences makes little or no reference to evolution. Is it because evolutionary theory has come to be so taken for granted that there is no incentive to question it or, more simply, because evolution has had little to do with the research framework behind the major developments in molecular biology that now inform most biomedical work?[21] In either case, it may be that public professions of faith in evolution by scientists are best interpreted as simply marking solidarity with the reigning orthodoxy. In any case, the alleged existence of a 'scientific consensus' provides the most rhetorically potent resource at the evolutionists' disposal to contest the legitimacy of ID. This then licenses a strategically biased framing of the issues at stake. Two recent cases stand out for their insulting clarity.

The first case is the InterAcademy Panel of 67 national academies of science, which in June 2006 issued a joint statement calling for the promotion of evolutionary theory in schools against an alleged worldwide creationist threat. Yet, the statement failed to identify any specific evolutionary mechanisms for mandatory instruction, resorting instead to an anodyne characterisation of evolution as somehow responsible for the diversity of life on Earth, and an easy dismissal of young-Earth creationism. One suspects that to manufacture the appearance of widespread scientific consensus, the statement's drafters were forced to edit out any specifically Darwinian references, especially to natural selection, that might prove divisive to some of the signatories.

The second case is the Council of Europe's condemnation of creationism as a potential threat to human rights, which was agreed by 47 countries by a 2–1 margin in October 2007. This openly polemical document played on the recurrent nightmare of a secular Europe beset by an ideological pincer attack from American Christian fundamentalists in the West and Asian (specifically Turkish) Muslim fundamentalists in the East. However, unlike the signatories of the InterAcademy Panel statement, it is unclear just how much the European parliamentarians understood of the science they supported. After all, it is odd that evolution would be portrayed as a bulwark of distinctly *human* rights against creationism, when it is the latter, not the former, that has historically privileged humans as possessors of a unique set of inalienable rights. In any case, the Council of Europe followed the InterAcademy Panel's strategy of conflating ID with other versions of creationism, but this time with conspiratorial overtones reminiscent of US Cold War intimations of hidden alliances between social democratic Europe and Soviet Communism, which of course also shared common ideological roots.

But perhaps the most paranoid appeal to a scientific consensus to defend against an impending Dark Age has been the promotion of the historically spurious idea that a specific metaphysical doctrine is embedded in the code of scientific conduct. Thus, the NAS would have scientists pledge a loyalty oath to a particular philosophical world view before being licensed to practice. This world view goes by the name of 'methodological naturalism', a phrase adopted from the wishfully named National Center for Science Education (NCSE), a California-based intellectual vigilante group that supplies pro-evolution propaganda at all levels of sophistication, contributes to the aggressively populist science blog, 'The Panda's Thumb' (nicknamed 'Darwin's Brownshirts' by its detractors) and, most notably, cooperates with the American Civil Liberties Union (ACLU) in its campaign to ban the teaching of creationism and ID in science classes. Specifically, the NCSE has on tap a battery of well-rehearsed expert scientific witnesses like Kenneth Miller, co-author of the best-selling US high school biology textbook, willing to testify under oath to the exclusive prerogative of Darwin's theory of evolution. Before turning to the creature of convenience that is methodological naturalism, a word about the ACLU is in order.

The ACLU came to national prominence in some high-profile cases in the 1920s, when it volunteered a quality of legal aid superior to that on offer to desperate defendants from local counsel. Among these cases was *State of Tennessee v. John T. Scopes* (1925) – the notorious 'Monkey Trial', the oldest and still iconic courtroom encounter between the forces of evolution and creation. That trial was not about disallowing the teaching of creationism but allowing the teaching of Darwin's theory of evolution, a scientifically promising but politically controversial account of the human condition that not only denied our divine origins but also appeared to rationalise racial inequalities. Indeed, the textbook used by Scopes, George William Hunter's *Civic Biology* (the title itself a euphemism for eugenics), presented primate evolution as a linear progression across the races, culminating in Caucasian supremacy. Keep in mind that the abolition of slavery was within the living memory of American senior citizens, not least the trial's special prosecutor William Jennings Bryan. Nevertheless, it was certainly within the civil liberties remit of the ACLU to defend the teaching of morally abhorrent yet scientifically supported doctrines in state-supported schools.

However, it is not clear why the ACLU should now be devoting its energies to banning ID, evolution's historically most robust alternative. In any case, the effectiveness of the ACLU's opposition cannot be denied. It has helped to preserve the illusion of a scientific consensus, now seemingly unified against ID. A key component here has been the warning shot that the ACLU fired to other potentially litigious dissenters by the $1 million settlement that the defendants in *Kitzmiller* were forced to pay. This was largely related to the surfeit of lawyers present daily at the trial, as well as extensive consultation with NCSE members who were also regularly present. Because *Kitzmiller* counts as a First Amendment case, under US law the plaintiffs can recover their attorneys' fees from the defendants, which in this case bankrupted the school board brought to trial. While few people's minds were changed by Judge Jones' verdict, the fear of bankruptcy has deterred other school boards from following Dover's example of introducing ID into the science curriculum.[22]

'Methodological naturalism', despite its philosophical-sounding name, has no clear meaning outside of attempts to demonstrate that creationism and ID are non-scientific. Professional philosophers, not least those who hold no brief for creationism, have squirmed at the apparent manufacture of a pseudo-doctrine customised to restrict the ranks of scientists.[23] This so-called principle conflates two 20th-century pro-science movements: 'logical positivism', which defined science in purely procedural terms as a method for testing theories, and 'metaphysical naturalism', which defined science as a world view that admits only causes like the ones already observed in nature. Logical positivists and naturalists overlapped in some but not all of their tenets: whereas logical positivists tended to be neutral on metaphysical issues, naturalists tended to be flexible about matters of testability.[24]

Each philosophy was responding to a different aspect of the history of science. On the one hand, positivists noted that scientists often used outlandish theoretical assumptions to reach empirically sound conclusions. In the current intellectual climate, many of these assumptions would be prejudicially dubbed 'supernatural', though philosophers would technically call them 'realist'. (When I was a student, they were called 'occult'.) They have included Newton's postulation of gravitational attraction and Mendel's of hereditary factors. On the other hand, naturalists observed that scientists have been in periodic power struggles with religious authorities over what counts as knowledge. Galileo and Darwin spring to mind in this context, though Darwin himself suffered from, if anything, celebrity rather than persecution. The two philosophies thus located the value of science rather differently. The logical positivists directed their arguments against all forms of unsupported authority in the spirit of keeping science as an open enterprise, whereas the naturalists specifically targeted religious authority in the spirit of replacing it with a superior scientific alternative.

Perhaps unsurprisingly, the scientific establishment has regarded both positivism and naturalism with suspicion, as evidenced by the prickliness of distinguished scientists when faced with sympathetically critical philosophers from either camp.[25] After all, positivists would dilute whatever advantage the dominant paradigm enjoys by saying that its explanatory power is only as good as its empirical reach, which in turn may be matched by other theories that explain the same things very differently. For their part, naturalists would force scientists, as guardians of epistemic authority, to spell out what their theories mean for everyday life. While some scientists with strong ideological views – say, ecologists, racists or atheists – relish this opportunity, most shun making normative pronouncements, lest their research be held hostage to political fortune. Indeed, what would likely happen to the research agenda of environmental science if the threat of global warming were to turn out to have been grossly exaggerated?

ID supporters treat appeals to the pseudo-philosophy of methodological naturalism as little more than anti-religious bigotry. I am reminded of the trumped-up 'brain-washing' charges associated with alleged Communist infiltration of the US government at the height of the McCarthy era, except instead of an alcoholic megalomaniac politician, now we have such academically certified luminaries as Richard Dawkins and Daniel Dennett calling family-and-church-based (as opposed to academically licensed) religious instruction 'child abuse'.[26] Indeed, there may come a time when sociologists will need to obtain special permission, beyond informed consent, before administering a survey that exposes subjects to various ways in which science and religion might be related to each other – out of fear that the mere exposure to alternatives might constitute indoctrination. Already there have been calls to rescind the natural science doctorates of those who deploy their scientific training to defend creationism in academic settings, as if that alone constituted professional malpractice.[27]

Such calls revisit the idea that a successful doctoral examination requires that one not only competently defend what one says but believe it as well. Just this additional requirement converted the medieval *inquisitio* from an academic exercise (i.e. the oral defence of the doctoral dissertation) to an instrument of religious persecution. Yet, arguably it was the removal of the requirement of personal belief that enabled modern experimental science to advance as rapidly as it has. Once conventions were established that allowed the testing of hypotheses under publicly observable conditions without prejudice to the hypothesiser, what mattered was no longer the confidence or even the sincerity of one's belief in the hypothesis, but simply one's ability to articulate the hypothesis in sufficiently explicit and plausible terms to establish its truth or falsity. Not surprisingly, reflecting on this history,

Karl Popper argued that belief is an altogether irrelevant category in science: what matters is not what you believe but what you can show, even to those who do not share your beliefs.

The bigotry lurking beneath methodological naturalism is ultimately fuelled by a paranoia regrettably familiar from the annals of US political history – except that now the ideological tables have been turned, with the left persecuting the right in the one arena where the left's dominance remains more or less assured despite eight years of George W. Bush: academia.[28] But the wages of bigotry are hard-earned. Public opinion surveys have consistently found that two-thirds of Americans would have some form of creationism or ID taught alongside evolution, up to 40 per cent would actually have creationism replace evolution and 40 per cent believe in divine intervention. When asked how these figures are possible in the world's leading scientific nation, thinkers like Dawkins and Dennett simply scratch their heads. Of course, it may simply show that one's position on the question of life's origins is irrelevant to most of the pressing questions of science – and hence can be safely debated without jeopardising the course of civilisation.

CHAPTER 2

Was Darwin Really a Scientist?

Science and Religion: Separate but Equal?

In the run-up to the 200th anniversary of the birth of Charles Darwin and the 150th anniversary of the publication of his masterwork, *On the Origin of Species*, both in 2009, the social standing of science is at a crossroads. But we have been here before. Take this excerpt from the judicial decision delivered in *Kitzmiller v. Dover Area School District* (2005), the first test case for the inclusion of intelligent design (ID) theory in US high school textbooks:

> After a searching review of the record and applicable case law, we find that while ID arguments may be true, a proposition on which the Court takes no position, ID is not science.[1]

The quote is striking *not* because Judge Jones disqualifies ID as science. That point can be easily understood in terms of the myth of a 'scientific consensus' that was exposed in the first chapter. Rather more disturbing is his suggestion that there are at least two truths, one for science and one for religion, with ID falling into the latter category as far as the judge was concerned. Had the judge not ruled in their favour, conformist pro-science thinkers would have dismissed his verdict as 'postmodern relativism' for endorsing the idea of multiple truths. Indeed, 400 years ago an analogous judgement had led the Roman Catholic Church to condemn Galileo, who dared to argue that only certain readings of the Bible, ones that reflected humanity's gradual realisation of the biblical message by scientific means, enabled access to the *one* divine truth.

In terms of US political history, the judge's declaration that science and religion constitute 'separate but equal' spheres of human experience – or what the palaeontologist Stephen Jay Gould dubbed 'dual non-overlapping magisteria'[2] – recalls the original justification for racial segregation that persisted in American law from Abraham Lincoln's formal emancipation of Black slaves in 1862 to Lyndon Johnson's Civil Rights Act of 1964. In both cases, the separation is a creature of political expedience. If this judgement seems harsh, then I wait with bated breath for the day when brain localisation studies make good on the judge's speculative psychology, which suggests that some non-question-begging sense of 'scientific' and 'religious' thought might be found in, say, opposing cerebral hemispheres. To be sure, an unholy alliance of cognitive anthropologists and evolutionary psychologists expect just such an epiphany through some combination of DNA divination and magnetic resonance imaging, the 21st century's answers to the astrologer's star charts.

So far, the legal case against ID has been easily made because of a precedent set by the US Supreme Court in *Edwards v. Aguillard* (1987), which interpreted the 'Establishment Clause' separating Church and state in the US constitution very broadly to imply that a form of inquiry explicitly motivated by religious concerns does not belong in publicly funded schools, regardless of how those concerns might substantively inform the subject in question. This interpretation is 'broad' because the original point of the Establishment Clause, as the name suggests, was merely to prevent the establishment of an official state religion. It was instituted in deliberate opposition to the situation in England, where an established Church had forced religious dissenters desiring a political voice to flee, resulting in the original North American colonies. It is easy for those of us living in the UK to forget this point, since for many years now the Church of England has occupied the left of the political spectrum.

In any case, the US Establishment Clause was not meant to prevent the airing of religiously inspired views in state-supported settings. The founding fathers would have marvelled at the perversity of *Edwards*. After all, they were the ones who came up with the idea of putting 'In God We Trust' on US currency – the ultimate symbolic yoking of the sacred and the secular. To be sure, few of the founders – certainly not Benjamin Franklin or Thomas Jefferson – could be counted as churchgoers, but they were all still spiritually aligned with Christianity and kept chapels in their homes for private worship and reflection.

Against the backdrop of religious wars between Catholics and Protestants that blighted Europe for most of the 16th and 17th centuries, the idea of the United States as a non-denominational 'Christian' nation under secular democratic rule was a remarkable innovation in political theology. The only previous model of a universal Christian community was the Roman Catholic Church, which imposed doctrinal unity by presuming that certain individuals, by virtue of their proximity to the papal office, had privileged access to God. Even the regionally based Protestant polities, while more open to secular democratic rule, required doctrinal unity on their particular versions of Christianity. In this respect, the United States was the first Christian nation that was constitutionally bound – albeit only in principle – to respect all human beings for their innate God-like powers rather than their conformity to a pretender to divine authority. Within a decade, the French would completely humanise the political rhetoric of the American Revolution, though without the institutional mechanisms to ensure that the decisions taken by divinely inspired free individuals would add up to a paradise on Earth.

Judge Jones' supposition that there are multiple truths – or at least a double truth, one for religion and one for science – was precisely what impeded the advancement of science until the modern era. The issue came to a head in the 1633 Vatican trial of Galileo, the iconic moment when science asserted its superiority to religion as a form of knowledge. Galileo's crime was not that his account of the heavens varied from that in the Bible. No, it was that he thought that the variance constituted a *contradiction*. He presumed that science and religion address the same divinely created reality, and found that scientific and religious authorities were saying things that could not both be right. Now, nearly four centuries later, a politically appointed Republican judge in Pennsylvania has handed the advantage back to Galileo's persecutors by reasserting the doctrine of double truth – except that the judge would have the doctrine administered by the scientific, as opposed to the religious, side of the divide.[3]

It is scandalous that historians, philosophers and sociologists of science have not been publicly forthcoming on a point that we easily grant in the classroom, namely that ID has provided the clearest justification for the respect and resources that science has enjoyed. The burden of proof lies with ID's Darwinist opponents who continue to deem science a worthwhile enterprise even if we are products of chance-based processes over which we are never likely to gain decisive control. Here we must be clear what is meant by 'science', which is a theoretical whole much greater than the sum of techniques needed to get by on a day-to-day basis, the *modus operandi* of every other species. All animals have sophisticated cognitive capacities for dealing intelligently with the physical world, not least by reconstructing their habitats in ways that enhance their chances for survival. Richard Dawkins has written of the 'extended phenotype' in this context.

To be sure, scientific progress is sometimes portrayed as the distinctly human extension of this general evolutionary proclivity. But this is to miss entirely the point of science, which is to do with a *unified* understanding of *all* reality, not just the specific bits that permit specific groups reproductive advantage. Science does not make sense unless reality possesses a *depth* that eludes our normal sensory encounters with the world but can nevertheless be accessed with sufficient application under the right conditions. Thus, science presupposes the *intelligibility* of reality; that its organisation, whatever its ultimate cause, is tractable to the human mind. Evolutionists have been much quicker to explain religion than science itself, yet it is the latter that should worry them, given the peculiar combination of mental dispositions needed to sustain scientific inquiry.

Science as a Precarious Evolutionary Product

If science's horizons were as limited as evolution suggests, a mere extended phenotype, then 'racial science' should be cognitively, if not morally, acceptable – not a politically incorrect oxymoron. After all, the idea that each form of life produces its own ecologically valid form of knowledge that enables successive generations to thrive in a common physical environment has a strong history independent of the boost it received from Nazi *Rassenwissenschaft* ('race science'). It underwrites time-honoured political and economic intuitions about inheritance as the securest form of societal reproduction. These intuitions were gradually extended into a general proposition about the biological rootedness of social life, starting with the introduction of *Kultur* ('culture') by Johann von Herder in the early 19th century, through Edward Westermarck's Darwin-inspired naturalistic conception of 'ethical relativity' in the early 20th century, to Jakob von Uexküll's 'biosemiotics', a prototype for today's ethology and evolutionary psychology. The Nazis simply turned this line of thought into a serious science policy in 1933, their first year in power, by liberating animals from the cages where they were held as experimental subjects. Some ethologists like Konrad Lorenz were impressed and remained loyal to the Nazi regime, while others, like Dawkins' teacher Niko Tinbergen, emigrated.[4]

While Darwin himself certainly saw science as a signature feature of our humanity, like latter-day evolutionists he said very little about its provenance or prospects. However, it is clear he held that the higher qualities of mind required for science, including a sense of progress, could be found in the lower animals to varying degrees. The overall picture that emerges from, say, *The Descent of Man* is that the more advanced intellectual achievements of humanity are precarious. First, Darwin stressed that relative to the instinctive proficiency of other animals, humans learn from experience slowly, each individual typically learning from scratch through imitation and practice. He was also inclined to explain cross-cultural similarities in artefacts and techniques in terms of independent invention by similarly-minded individuals, rather than the diffusion of accumulated knowledge. (In this respect, Darwin appeared to underestimate the extent to which modern science itself has been a 'made for export' Western product.) The inherent limits of knowledge transmission extended, in his view, even to the potential hereditary benefits of education: Darwin believed that education was wasted on women once they bore children. Based on the anthropological evidence provided mainly by John Lubbock, Darwin believed that innovations emerged accidentally and required a materially favourable environment for their cultivation, which itself could be easily upset by a surplus of incompetent offspring. Like his nephew Francis Galton, Darwin hoped, but did not necessarily expect, that legislators would someday modernise marriage laws to reflect these eugenic concerns in order to sustain human progress. But perhaps most tellingly, Darwin believed that the propensity for strange and maladaptive behaviours, a product of the limits of the animal mind, was magnified in humans as involuted abstractions were promoted to fixed ideas and superstitions, only to result in much useless violence and warfare; what Darwin epitomised as 'barbarism'. Thus, Darwin strikingly ends *The Descent of Man* with the observation that he feels better knowing that he descended from baboons rather than from the warriors he encountered in Tierra del Fuego at the tip of South America while sailing on the HMS *Beagle*.[5]

In short, according to Darwin, we are lucky to have got as far as we have: there is no guarantee that we shall improve indefinitely, let alone ever be in a position to take control of evolution, as ID traditionally promised. Under the circumstances, species humility – if not species egalitarianism – is the only reasonable attitude to adopt. Whereas his contemporaries John Stuart Mill and even T.H. Huxley, both staunch anti-clericalists, nevertheless still retained a faith in humanity's capacity for global domination, Darwin's political horizons were closer to those of his fellow evolutionist Herbert Spencer, who expressed general unease with, and sometimes outright opposition to, the British Empire's radical reorganisation of distant native practices. Indeed, while Darwin is rightly cited for intervening in the early 1870s to quell the anti-vivisectionists who were trying to shut down the physiology labs, it is clear that he believed that animal suffering was justified only insofar as the resulting research was likely to reduce further suffering – a utilitarian argument now associated with the philosopher of animal liberation

Peter Singer.[6] Given that medical science's modern aspirations have generally included the extension of life and the enhancement of our powers, Darwin's support for medical research would have to be seen as highly qualified. All in all, if Darwin suddenly appeared today, equipped only with the understanding of the world he had upon his death in 1882, he would immediately find kindred spirits among those environmental campaigners who doubt that global warming will be ultimately overcome through the sort of 'technological fix' that the design perspective has tended to encourage.

Darwin's own scepticism notwithstanding, modern science's drive for a rationalised world order has motivated and justified the systematic ecological transformation of the Earth, for reasons that only start to make sense once we abandon the logic of selective reproductive advantage for the broadest range of species. I mean to include here such developments as the mass eradication of environments that breed pathogenic organisms, the escalation in greenhouse gas emissions, and 'Green Revolution'-style agricultural innovations. All three have arguably reduced the world's biodiversity and contributed to a seemingly never-ending series of climate-based crises. Such policies, which collectively used to be called 'development' (as opposed to 'evolution'), focus on the promotion of human welfare above all else. Science can justify this goal only by assuming that we are unique in our comprehensive grasp of nature, which in turn entitles us to govern the planet for our collective benefit. The obvious historical source for these two assumptions is the Bible.

In this respect, science's default attitude towards nature is well exemplified by imperialism in both its capitalist and socialist phases, and its latter-day descendant, globalisation. The planet's mode of production has been reconfigured to be efficiently exploited by *Homo sapiens* in its guise of *Homo faber*, 'man the maker', on the assumption that every new problem unwittingly generated by this reconfiguration will be solved by the appliance of still more science. Even Marxists have been much less concerned with imperialism's evisceration of nature than the fact that most humans could not participate fully in the process, as long as their own labour remained exploited as if it too were a generic part of nature. For Darwin, this entire line of thought, which has animated the political and economic imagination for the last 150 years, should have by now led to global misery, if not the extinction of *Homo sapiens*. But it has not, though there is certainly room for reservations, especially given humanity's increased capacity for strategic self-destruction. In any case, such naked anthropocentrism merely embodies and secularises the optimistic spirit that lay behind the God-intoxicated invention of the 'scientist' as a distinct social role in Darwin's own day.[7]

Darwin never regarded himself as a 'scientist' – though not for reasons of personal modesty. True, he saw his powers of reasoning as being no greater than those of the average lawyer or doctor, though perhaps with a keener eye for the telling detail. Indeed, he expressed amazement that 'men of science' had made so much of his work.[8] However, behind this appearance of modesty lies tact. The word 'scientist' was coined only in 1831, the year Darwin began his famous voyage on the *Beagle* which established his reputation as a naturalist. The word referenced a divinely inspired attitude to nature that by the publication of *On the Origin of Species* he had come to disown as epistemologically untenable. However, *Origin*'s epigraph kept the illusion alive a bit longer. It came from the man who coined 'scientist', William Whewell, Master of Trinity College, Cambridge, proficient in both natural philosophy and natural theology, and a founder of the British Association for the Advancement of Science.

Before Whewell, 'science' referred to a systematic body of knowledge that might be pursued by anyone with sufficient leisure, as in the case of Darwin himself. In contrast, the idea of 'scientist' suggested someone who devoted his entire life to science after having undergone a disciplinary regime akin to that of the priest or a monk – 'science as a vocation', in Max Weber's memorable words. However, such a single-minded dedication to science would not make sense without faith in the intelligibility of all nature. Otherwise, humans would more wisely spend their time on more modest attempts to ameliorate our transient animal existence. Darwin began writing *Origin* with Whewellian ambitions, which first drew readers to his masterwork; namely, the diligence with which Darwin synthesised the observations of several naturalists of many species across the world. However, the prominence given to a purposeless force of natural selection in evolution signalled that Darwin had come to regard the career path formally named by Whewell as futile, if not altogether chimerical: there was simply no unity to be found in nature.

Whewell's faith in lifelong dedication to science was buoyed by the example of Isaac Newton, who leveraged a biblical understanding of humanity's cosmic privilege into the broadest and deepest scientific theory the world has ever known. Like other figures associated with the Scientific Revolution of the 17th century, Newton took literally the idea that the universe is a divine artefact; specifically, a great machine whose design we can reverse-engineer, and possibly improve and even perfect. This attitude continued to inform the scientists who migrated from physics to biology in the 20th century, first into genetics and then molecular biology, to such an extent that biotechnology forms the vanguard of today's life sciences. It is a line of thought completely alien to those who took the traditional route into biology via natural history and field studies, outside an explicitly agricultural context. These figures, Darwin included, treated life as ultimately an irrational force of nature. Unlike his contemporary who is now credited with the foundation of modern genetics, the Catholic monk Gregor Mendel, Darwin stressed the *dissimilarities* between the overriding process of 'natural selection' and the familiar techniques of 'artificial selection' used to breed plants and animals. For Darwin, 'natural selection' was, as we now put it, a 'science-stopper' that provided an absolute limit to our comprehension and control.

That we deem 'scientific' a chance-based theory of biological evolution that specifically excludes ID simply reflects the inordinate influence of Charles Darwin's personal history on the history of science. His initial captivation by the idea of ID motivated a lifelong intensive study of organisms, the conclusions of which nevertheless failed to vindicate his original motivation. As it happens, Darwin's course to a godless universe was originally interpreted not as a case of the theistic scales falling from his empiricist eyes, but as indicative of a depressive personality whose assessment of the evidence for mass extinctions from natural history was distorted by senseless death in his own family. Under this interpretation, popular among early-20th-century theologians looking for God in evolution, Darwin's genius lay in just how much he had managed to see *in spite of* his psychic liabilities.[9] A secular version of this spin is routinely put on the life of Newton, whose genius somehow managed to overcome his theistically warped world view. Thus, at the time of writing, a museum exhibition relating to Darwin's life, including his domestic arrangements, is circulating worldwide, whereas nothing of the sort took place for the 300th anniversary of Newton's *Principia Mathematica* in 1987. The public relations implication is clear: the more one knows about Darwin and the less about Newton, the more sympathetic both appear.

Darwin's uniqueness lies less in the conclusions he reached than the path he took. Most naturalists who had come within the vicinity of Darwin's theory of evolution by natural selection, such as the Epicureans in the Greco-Roman world and the various karmic cosmologists of the East, possessed much less evidence, often relying on no more than first principles combined with personal introspection and a sympathetic understanding of other life forms. Unsurprisingly, their testimony was much less persuasive in monotheistic lands because they failed to do justice to the apparently designed features of nature. Darwin, of course, did ID justice – before rejecting it. However, for those who accept Darwin's negative verdict, the question it raises is why *they* should continue studying nature as intensely as Darwin did. The answer, of course, is that most contemporary biological research is not beholden to Darwin's purposeless vision of life. The non-Darwinian history of modern biology, which goes from genetics to molecular biology to biotechnology, certainly vindicates the idea that nature has been designed with sufficient intelligence to be susceptible to purposeful human modification. This is a conclusion worthy of the title of 'science', something that Darwin once again claimed *not* to have practised. I shall return to this point in chapter 6.

Science as the Natural Outgrowth of Monotheism

Given his own mysterious account of life's origins and incoherent understanding of heredity, Darwin would be amazed to learn of the fine-grained detail with which we have come to both understand and manipulate the fundamental stuff of life. If anything vindicates Whewell's supposition that the universe has been created in a manner tractable to the human mind, this must certainly be it. It is a supposition shared by today's ID theorists.

The idea that science can be properly done only within a monotheistic framework is traceable to the theologian who played a large part in the creation of 'Church Latin' in order to stop the Christian message from being diluted by foreign tongues. Tertullian, writing in AD 200, when the Bible's contents were still very much up for grabs, provocatively asked: 'What has Athens to do with Jerusalem?' He ridiculed the founder of Greek science and philosophy, Thales, who, as court astrologer to the Egyptian Pharaoh, reportedly fell into a ditch because his attention was so fixed on the heavens. Tertullian snorted that Thales deserved his fate because he should have been focused on fathoming the divine creator rather than indulging his curiosity in the passage of creation.[10] In these secular times, it is natural to read Tertullian's scorn as drawing the faithful from the study of nature to the study of the Bible: choose religion *over* science. In fact, Tertullian meant that nature can be studied properly only *through* the Bible. Unfortunately the spirit of his injunction has been lost, even by those who see themselves as sympathetic to it. Before we can make headway through the current impasse in the evolution–creation debates, we need to cast Tertullian in the right light.

Consider Thales once again. To our eyes he looks like an absent-minded scientist more concerned with the objects of his inquiry than what lies beneath his feet. We excuse his behaviour because we have no trouble envisaging that the ultimate nature of things may be radically different from what it takes to get by on a day-to-day basis. The main thing to understand about Tertullian is that he adamantly denied this interpretation of Thales' situation: according to him, a genuine focus on the ultimate nature of things should enrich, not detract from, the conduct of everyday life. Tertullian read Thales' upturned head to symbolise not high-minded pursuits, but base attraction to the shiny specks in the sky that happened to catch his fancy. Here 'curiosity' should be understood *negatively*, as 'idle curiosity', an animal passion that potentially undermines a systematic understanding of divine creation by granting the senses too much authority over what counts as knowledge worth having.[11]

If curiosity did underwrite the scientific imagination as evolutionists often presume, then science would be indistinguishable from organised nosiness, gossip, prurience or perhaps even the spectrum of attention-deficit disorders that command an increasing portion of biomedical funding. To Tertullian – but hopefully not just him – this would be problematic, as the casual invasiveness of the prying mind demeans both the observer and the observed. In contrast, the systematic pursuit of science ennobles both by requiring a principled commitment to understand people and things in their own terms – as enactments of the divine word, or *logos* – and not simply in terms of the temporary satisfaction with which they provide the observer. How one first acquires and then uses such knowledge of people and things 'in their own terms' are, of course, open questions. For example, an important reason for wanting to so understand people and things is to grasp their *modus operandi*. This might then make them easier to manipulate, an entitlement that the Bible extends to humans over all creatures other than fellow humans.

To be sure, even on these terms it has been historically easy to restrict the autonomy of scientific inquiry. Consider the development of science in medieval Islam. It was the first culture explicitly committed to the universalist ideal that all genuine knowledge is a human birthright. Islam is the source of the very idea that pagan Greco-Roman learning should be systematically reconciled with sacred scripture – rather than adopted selectively, rejected outright or simply treated as a curiosity. Nevertheless, by our lights, medieval Islam interpreted the development of this collective intellectual endowment in increasingly conservative terms: technologies were improved without allowing the questioning of fundamental scientific principles. This situation tends to be portrayed, unfairly in my view, as exemplifying medieval Islam's prohibition on free-ranging curiosity, as if greater Muslim curiosity would have issued in more profound scientific breakthroughs. No, free-ranging curiosity would have simply fragmented the collective knowledge base, as inquirers followed their particular fascinations. The truth is probably much more prosaic: medieval Islam gave disproportionate weight to applied over basic research because people with sacred (in our sense, 'scientific') authority came to be invested with secular power, which allowed the long-term vision instilled in them through their training to be compromised by the short-term vision required by their office. In this respect, Christendom's saving grace has been the relative difficulty with which the sacred has been translated into the secular. While science is undoubtedly powerful, scientists have rarely exercised power, finding themselves more often in the ranks of heretics.[12]

In any case, Muslims and Christians – and Jews – were agreed that the self-discipline required for science did not imply a passive observance of nature. On the contrary, the descriptivist mode preferred by Darwin and other naturalists down through the ages has been regarded with considerable suspicion. Notwithstanding the divine pedigree of humanity, only laziness cloaked in arrogance would let us conclude that whatever patterns in nature attract the human senses are in fact those that are most significant to God. Reality is deeper than that, a lesson that Tertullian would have taught Thales. Rather, a peculiar sort of humility is required, one that has made experimentation central to the scientific method. After all, experiments make sense only if one suspects that appearances are literally not all that they seem – but are no less significant for that. But why be suspicious at all, unless one takes seriously the idea that reality possesses a deep yet fathomable plan? While theological opinion has warned that excessive fascination with the depiction of animals verges on the sin of idolatry, it has equally supported the study of animals as spurs to human ingenuity that enable us to bring creation to full realisation. Thus, scientific supporters of ID have been attracted to biology's interfaces with physics and engineering. Such fields as 'bionics' and 'biomimetics' proceed from the premise that organisms (possibly as divine artefacts) throw up prototypes from which humans may learn and which we may refine for our own purposes, and that this gives a direction to nature that it might otherwise lack. I shall develop this point in the next chapter.

Science's activist, even constructivist, orientation to nature suggests a larger point. While scientists, especially when on the defensive, make a big rhetorical fuss of how seriously they take evidence, the most salient feature of the scientific enterprise is that it is *theory-driven*. This means that scientists begin with a certain view of the world, which they then test against the evidence, in light of which their view is then revised and extended accordingly. Thus, philosophers of science used to say that science is governed by something called the 'hypothetico-deductive method', which brought out clearly the idea that theories are speculations whose merits are judged against their derivable consequences, which are treated as predictions. Today this way of speaking sounds old-fashioned, and to some even misleading, but it retains a certain force: why *do* scientists proceed in this rather indirect way, rather than in a more immediately responsive ('inductive') mode to nature? The latter seems to be more in line with what evolution would recommend as the best survival strategy; namely, to produce knowledge that enhances our capacity to adapt to the environment. To be sure, much of what passes for science does have this character: durable housing and clothing, transportation and communication infrastructures, health delivery systems, standardised weights and measures. But in these cases, the reason for labelling such responsive activities 'science' rather than, say, 'technology', 'engineering', 'medicine', 'business' or even 'politics' becomes unclear. This is a point repeatedly made by researchers in the controversial field of STS.

As we saw in the first chapter, STS researchers are frequently accused of being anti-science, but a fairer description is that they are 'anti-' the *concept* of science as a tool for understanding scientists and their activities. In other words, STS researchers do not question the actual results of scientific inquiry, only the larger significance ascribed to them: what licenses extrapolations from the lab or the field to the world at large? It is here – in the realm that methodologists call 'validity' – that the concept of science looms large and mysterious. In this sense, STS researchers are rightly called 'demystifiers'. STS research has shown some empirically straightforward ways of making the world conform to the lab or the field. But it is here that science most clearly turns into technology, engineering, medicine, business and/or politics: additional constraints, incentives and conduits are introduced to ensure that the world behaves in accordance with the findings. The interesting point for STS researchers is when these *ad hoc* manoeuvres start to appear strained, or 'artificial' in a negative sense, leading to the conclusion that maybe the findings themselves are at fault.

In contrast, the interesting question for me in these pages is almost the exact opposite: what is at stake in continuing to call such an activity 'science' rather than 'technology' etc., especially in the face of resistance from the new environments in which they are applied? Often the rhetorical import of calling something 'scientific' is to sign a blank cheque to reconstruct the world in the image and likeness of our theories – something that more specific appeals to mere 'technology' etc. would not permit so easily. This is because, I submit, despite a somewhat dubious etymology of the Greek *theos* that links 'theory' and 'theology' (and 'theatre', a point whose relevance will become clear in chapter 7 when discussing biblical literalism), there is nevertheless an historically deep connection between theological inquiry and scientific theorising: *scientists can only make sense of a world they could have created.*[13]

There is a modest and a bold way of taking this statement. Both are traceable to St Augustine, the 4th-century Bishop of Hippo in modern-day Tunisia, who remains the most influential philosophical interpreter of Christian doctrine. The modest way is the older and perhaps more natural reading: since we did not create the universe, we shall never fully understand its nature through reason alone. Faith is thus a prerequisite to comprehension; beyond a certain point, humans cannot think for themselves but must delegate their capacity for thought to God, such that he is allowed to think his thoughts through us as vehicles of his creation. In the previous chapter, I presented this reading of Augustine as a religious precursor to the Pasteurian maxim: 'Discovery favours the prepared mind.'

But Augustine can be also read boldly as drawing attention to the univocal character of 'creation': divine creation is indeed infinitely greater than human creation, but in both cases 'creation' means the same thing. God just does it much, much better; 'infinitely greater' means 'indefinitely extended', not something altogether unfathomable. This reading of Augustine was championed by the 14th-century British scholastic, John Duns Scotus, whose reputation as Christendom's 'subtle doctor' rested on his criticism of Thomas Aquinas for trying to shore up papal power by mystifying the difference between divine and human mental capacities. Duns Scotus' bold Augustinianism underwrote the peculiar combination of biblical literalism and intellectual adventurousness that culminated in the Scientific Revolution – and which we shall subsequently see evidenced in Gregor Mendel's foundational work in genetics. We can understand creation precisely because we are creators in the same sense as God: we organise matter into meaningful objects, just as God does. Without this assumption, it would be difficult to grant any *prima facie* plausibility to the idea that nature can be understood as a gigantic machine, as the original scientific revolutionaries did, let alone that a living cell can be understood as a mousetrap, as ID theorists do today. To be sure, our attempt to do as God has may always remain fallible and incomplete, but it is never beside the point. Reason may thus make considerable headway if propelled by a steadfast will.[14]

We shall have occasion in these pages to explore the seemingly jarring juxtaposition of biblical literalism and scientific boldness. But for now, consider a paragraph from Augustine's classic work on the meaning of Genesis that is often cited as evidence that he did not believe that the Bible was a scientific work. Interestingly, the paragraph's final sentence, italicised below, tends to be omitted:

> In matters that are obscure and far beyond our vision, even in such as we may find treated in Holy Scripture, different interpretations are sometimes possible without prejudice to the faith we have received. In such a case, we should not rush in headlong and so firmly take our stand on one side that, if further progress in the search of truth justly undermines this position, we too fall with it. *That would be to battle not for the teaching of Holy Scripture but for our own, wishing its teaching to conform to ours, whereas we ought to wish ours to conform to that of Sacred Scripture.*[15]

The final sentence makes it clear that Augustine is cautioning against not literalism *per se* but the conflation of knee-jerk readings of the Bible with its literal meaning. Much of Augustine's tract in fact concerns errors of interpretation that arise from mistaken translations and misunderstood contexts. Bearing in mind that most readers of the Bible until the modern era were likely to have been involved in its transcription, the inevitability of corruption across great expanses of time and space demanded that each reader make sense of the text for him or herself, testing one's beliefs against it, not the other way round. This became the signature attitude of the Protestant Reformation and, indeed, the Scientific Revolution, as in the Royal Society's motto *Nullius in verba*, 'On the word of no one.' We shall encounter it below in Francis Bacon's famous metaphor, inspired by Galileo, of Holy Scripture and the natural world as two modes of inscribing the same common reality.

Linnaeus and Vico: Enlightenment Science in the Biblical Mode

The person whose scientific practice perhaps most consistently exemplified Augustine's strong sense of literalism was the 18th-century Swedish physician-turned-naturalist, Carolus Linnaeus, to whom we owe the binomial system of classification by which species continue to be identified – as in *Homo sapiens*, one of his coinages. Linnaeus was a 'special creationist', for whom the concept of species captured the immutable properties that make God's creatures what they are. Linnaeus' now-familiar taxonomy of living things into kingdoms, orders, genera and species was tantamount to a caste theory of nature, with the chance for mobility – that is, species change – restricted to the hybrids created through animal and plant husbandry, thereby allowing at least a modest role for humans in the project of divine creation.

However, Linnaeus also had larger ambitions, believing that his taxonomy elucidated the method that Adam originally used to name the creatures in the Garden of Eden. On this point Genesis is elliptical, perhaps presuming that readers could fill out the scheme for themselves. But Linnaeus realised that over centuries of transmission and translation, the original intent of the biblical authors had probably been corrupted, if not completely lost. Thus, the Adamic moment needed to be revisited with fresh eyes. The literalist cast of Linnaeus' mind should not be underestimated. Living in a time when stars and elements are numbered rather than named, we easily overlook the diligence required to systematically assign names to 4,400 animals, 7,700 plants and 500 minerals. This is a task that would probably not be undertaken today, since our post-biblical culture tends not to see meaningful names as containing power over those so named.

In Linnaeus' case, however, the two-name system applied uniformly across the three kingdoms – of animals, plants and minerals – reflected the idea that Adam's exercise was informed by a *logos* emanating from God that identified a function corresponding to each species as part of an overall plan. Thus, Linnaeus' masterwork bears the title *The System of Nature*; the names given to creatures did not form a mere list, as in a medieval encyclopaedia, but a grammar with rules for constructing new names. Moreover, the Linnaean grammar was an instruction manual for controlling nature by reading intelligent design from morphology, as in the case of *Homo sapiens* ('man the wise'), based on the distinctiveness of humanity's high forehead among the apes. To this day, many species retain the rather anthropomorphic names that Linnaeus gave them, reinforcing the biblical idea that nature was created *for* humans. Drawing on his original training as a physician, Linnaeus believed that the disorders that humans periodically suffer at the hands of nature resulted from creatures being addressed by the wrong names, which then led humans to deal with them inappropriately.

At play here is clearly more than the musings of a pious Christian who met the Enlightenment only half-way. Linnaeus' facility with names enabled him, much to the marvel of his patrons, to classify any form of life, even when he had never previously encountered the creature. This was a feat biblically associated with Solomon's wisdom (1 Kings 4:33). Linnaeus' trick was to examine the creature's sexual organs, which revealed a sense of human dominion over nature. The organisation of life forms by sexual characteristics, while not completely successful, highlighted Linnaeus' conception of organisms as machines designed to reproduce and ultimately replace themselves. Indeed, true to the original science of design, which as we shall see in chapter 5 was a theory of divine justice (theodicy), Linnaeus believed in a literal 'economy of nature' whereby births and death are always perfectly balanced in what we would now call the global ecology, a concrete sense of which was beginning to be gleaned in his lifetime from the increasing use of long-haul voyages for commercial, diplomatic and military purposes.

Linnaeus is usually portrayed as part of the prehistory of modern biology. But he is better seen, like his great contemporary and rival the French Count Buffon, as an early political economist of the generation prior to François Quesnay and Adam Smith, who are normally cast as the first proper economists. It is easy to forget that 18th-century natural history and political economy shared a preoccupation with the transformations of the Earth associated with generating and maintaining life. It is worth recalling that the words 'biology' and 'sociology' came to mean different forms of inquiry – and mainly in France – only in the first third of the 19th century. Before that time, and long afterward elsewhere, the transition between natural and civil history, as well as natural and political economy, was largely seamless. For example, Smith probably came to believe that the supply and demand for goods tend toward equilibrium from Linnaeus' observations about the finite size of animal stomachs checking an otherwise potentially insatiable desire for nourishment.[16]

Moreover, the economists of Smith's generation took particular notice of Linnaeus' arguments for national economic self-sufficiency, what at the time was called 'cameralism' but nowadays is better known as 'protectionism'.[17] Linnaeus dreamt that the emerging nation-states of Europe – and possibly elsewhere – might turn themselves into so many modern Gardens of Eden by containing within their borders all of God's creatures, the successful governance of which would enable humanity to recover from its fallen status. According to Linnaeus, Adam's command over all the basic types – if not the exact species – of organisms available today was the high-water mark of our closeness to God. Once each rationally self-constituting human society returns to that original state, he surmised, we shall seal our superiority over the brutes, which we resemble in nearly all other physical respects; this, in turn, explains our resort to such brutish actions as threats, theft, deceit and begging in the name of 'free trade' and 'diplomacy'.[18] Of course, Linnaeus did not quite get his way, but the spirit of his project lives on in every director of a zoological or botanical garden who frets over the comprehensiveness of his or her institution's holdings. Nowadays we treat zoos and gardens as places to see creatures that normally live in other parts of the world, but to their Linnaean founders they are temples to our mastery of nature, demonstrating that naturally alien creatures can live harmoniously in human-designed environments.

By his death in 1778, Linnaeus was widely regarded as the most significant scientist since Newton. Although the two differed in both topic and method, there was no denying that their careers exemplified a synergy between cutting-edge science and enlightened Christianity. Science continued to be indebted to creationism in the 19th century, and often in the most unsuspecting quarters. Thus, while Darwin denied that biological species were either specially created, immutable or intelligently designed, he nevertheless retained Linnaeus' binomial classification of species. Indeed, in his own later works and those of his followers, Darwin took another step back to Genesis by reviving the 'tree of life' image, which had played no role in Linnaeus' work. For his part, Linnaeus originally organised the species to look like what we would call a periodic table of elements, consisting of parallel hierarchically arranged patterns, which enabled him to account for *all* of natural history – both biological and geological formations – within a special creationist natural theology indifferent to historical relationships among the species.

In contrast, the arboreal image presented life as having descended from a common ancestor, through the action of a common principle (natural selection), to produce the manifold species in nature. Darwin himself initially took the idea of common descent as being quite speculative:

> Therefore, on the principle of natural selection with divergence of character, it does not seem incredible that, from some such low and intermediate form, both animals and plants may have been developed;

> and, if we admit this, we must likewise admit that all the organic beings which have ever lived on this earth may be descended from some one primordial form. But this inference is chiefly grounded on analogy, and it is immaterial whether or not it be accepted. No doubt it is possible ... that at the first commencement of life many different forms were evolved; but if so, we may conclude that only a very few have left modified descendants. For, as I have recently remarked in regard to the members of each great class, such as the Vertebrata, Articulata, &c., we have distinct evidence in their embryological, homologous and rudimentary structures, that within each class all are descended from a single progenitor.[19]

However, August Schleicher, the German scholar who introduced the tree-like structure, or 'cladogram', to the historical study of languages, contacted Darwin after reading *Origin of Species* to urge him to adopt the device in subsequent works, which Darwin did.[20] Inspired partly by the genealogies of the patriarchs and the accounts of the dispersal of the Earth's peoples provided in Genesis, Schleicher constructed a genealogy of Indo-European languages from an Adam-like originating moment in Northern India. While now widely criticised within linguistics for its Eurocentrism, Schleicher's cladogram has stuck in biology, greatly aiding in the development of taxonomy, especially once DNA came to be interpreted as coding moments of speciation, or branching of the evolutionary tree. This is the so-called 'molecular clock' hypothesis, to be discussed in chapter 4.[21]

Notwithstanding Linnaeus' religiously driven scientific career and Darwin's subtle creationist concessions, modernity's passage from the modest to the bold interpretation of humanity's biblical entitlement has been far from straightforward. In effect, a compromise has been struck, courtesy of another 18th-century figure, Giambattista Vico, the Neapolitan jurist who effectively split the difference between the two readings of St Augustine. Vico enjoys the dubious distinction of having done the most to promote the Catholic 'double truth' doctrine in secular intellectual life. He is usually credited with having laid the foundations for a controversial yet enduring division of the sciences into those of the spirit and those of nature, in which the former pertain exclusively to human creations. These may be known intimately, because we can imagine what it would be like to have done and said as other humans did and said. The latter pertain to creations in which humans did not originally play a role. They can be understood only in terms of how they appear to our senses, their inner workings hidden to all but the deity who created them. Vico's explicit target was René Descartes but might as well have been Newton. Both Descartes and Newton presumed that they could adopt the divine standpoint by proposing a simple set of underlying principles from which they claimed they could infer all the phenomena of nature. For the devoutly Catholic Vico, while it was perfectly acceptable to recreate in one's own mind the reasoning behind a human achievement, even one in the remote past, it was downright blasphemous to attempt to recreate the workings of something that first emanated from the mind of God.[22]

It would be easy to underestimate the long-term significance of Vico's strictures, even long after their theological inspiration has disappeared. Consider the ease with which we allow for speculative explanations in the humanities and social sciences, while insisting on a rather limited sense of 'empirical evidence' before licensing theories in the natural sciences. Our intuitive closeness to the phenomena studied by the 'human sciences' serves to lower the evidentiary bar, giving freer rein to the imagination, whereas our intuitive remoteness from the phenomena studied by the natural sciences raises the bar, effectively casting aspersions on whatever contribution our imagination might make to a systematic understanding of them. Vico himself concluded that we can know more about ourselves than nature because we are literally self-created and hence entitled to use our full being – including our imaginations – in the conduct of inquiry. At the same time, this attitude also helps to explain the air of suspicion that continues to hang over the word 'theory' in the natural sciences, with its hint of 'mereness', as if a theory were, at best, a summary of or, at worst, a substitute for the phenomena under investigation.

The Scientific Method, or the Divine Spark Disciplined

To speak as we have of the inherently theory-driven character of science is to push the bold reading of Augustine's maxim to the limit. A theory, in that case, becomes the simulated execution of some prior plan. To be sure, our grasp of this pre-existent design is usually flawed, often fatally, which explains the periodic need to change theories. But why even bother to engage with the world so assertively in the first place, if one already anticipates failure? After all, so much philosophical wisdom – albeit much more in the oriental than the occidental traditions – consists of counsel about the futility of effort and the need to minimise suffering. Yet scientists persist in courting, at an intellectual level, error and, at a material level, disaster. Why?

Impolitic though it may be to admit, to view science as an endeavour whose value surpasses that of other secular activities makes sense only if there is an overall design to nature that we are especially well-equipped to fathom, even though most of it has little bearing on our day-to-day animal survival. Humanity's creation in the image and likeness of God, a doctrine foundational to the three great monotheistic religions – Judaism, Christianity and Islam – provides the clearest historical rationale for the rather specialised expenditure of effort associated with science. The much-vaunted 'creative' dimension of science that culminates in 'genius' comes close to acknowledging this divine spark. Art differs from science only in the modesty of its ambitions, since science presumes a universality of scope that art does not. In other words, artists do not presume that we can all inhabit their conceptions, but that we should respect them nonetheless as reflective of the same (divine) creative impulse. In contrast, scientists would hold themselves – and us – to a higher standard, the ultimate convergence of human and divine intelligence, a prospect that certainly animated Newton, as well as his theologically and scientifically more ambitious followers such as the polymath chemical philosopher Joseph Priestley (about whom more below) and probably Whewell himself.

That Whewell had no qualms about the free play of the imagination in scientific inquiry is exemplified in his support of studies into the *aether*, the medium that many 18th- and 19th- century scientists believed was necessary for the conduct of light and other fundamental physical forces like electricity, magnetism and gravity. According to Whewell, science disciplines our spontaneous intuition of the supernatural into theoretical entities, by inspiring experiments that test for alternative models of the hidden mechanisms responsible for an array of effects that appear to us as light, electricity etc. Such mechanisms are justifiably called 'supernatural' because they are normally experienced only indirectly in terms of their effects, but not as they are in themselves.[23] In this respect Whewell followed Francis Bacon, Lord Chancellor of England in the early 17th century, who is popularly credited with the invention of the scientific method.

Bacon held that experimentation is not merely empirical observation by more exact means but a portal to a deeper sense of reality, the causal order of nature. As personal lawyer to King James I, he wanted to distinguish those who had gained legitimate access to the supernatural via experimentation from those, mostly astrologers and alchemists, who illegitimately capitalised on people's gullibility to claim knowledge they did not truly possess. Bacon's supernaturalist aim has been subsequently secularised as 'the real', whose philosophy is 'realism'. Bacon's approach to science reflected neither anti-religious nor even anti-theological sentiment, but the Protestant independence of his patron, which is commemorated whenever one reads the King James Version of the Holy Bible. Divorced from the Papal Office on matters of faith, Bacon was empowered to propose a legal regime to decide the state's intellectual direction. In response, he adapted for secular courts the Church of Rome's inquisitions, which examined witnesses under 'strict' conditions – including torture – to get behind their testimony's *prima facie* plausibility to reveal its true motivation. This meant, in practice, that the implications of one's hypotheses could be detached from one's state of mind, the latter accountable only to God but the former tested in open view. Thus, under Baconian duress, the alchemist or astrologer would be forced to translate his rhetorical extravagances into specific mechanisms responsible for specific effects under specific conditions that could be demonstrated without undue reliance on the personality of the adept. Thus was born the modern scientific method.[24]

The capstone of modern Western philosophy, Immanuel Kant's *Critique of Pure Reason*, is dedicated to Bacon. It is through Kant that Whewell understood Bacon.[25] Kant saw a strong analogy between the scientific and moral spheres: both realms are governed by laws that are in some sense self-legislated. In humans, this is experienced as an inviolate moral autonomy, whereby responsibility for our fate ultimately rests on the freedom we exert over our lives. But in each individual the scope of this freedom is constrained by allowing to oneself only what we would allow to everyone. (Liberalism and socialism in the modern era were founded on different interpretations of this principle.) Similarly, natural entities also contain their own potential for action, what Leibniz originally called *vis viva* ('living force'), but which by the 19th century had come to be named by such words as 'force', 'power' and, most popularly, 'energy'. Here, too, nature must be addressed in a suitably uniform fashion – using equations, making measurements and performing calculations – according to a common standard that also accounts for ourselves as natural entities. In this respect, the laws of physics codify the universe's sense of procedural justice. Even today, when we speak of the need to 'harness' energy, we concede to nature a mind of its own, whose terms we need to respect in order to get whatever we might want from it.[26]

Bacon memorably presented this dual aspect of divine creation – the will of humanity and the power of nature – as epitomised in 'two books' that equally merited serious study, the former by theologians and the latter by those proto-scientists, the natural philosophers.[27] In these secular times, we tend to stress the separateness of the two books. But this is to overlook the truly radical nature of the idea that the Holy Scriptures and physical reality can be treated as 'books' at all. In other words, both can be understood not as objects of awe that ultimately transcend human comprehension, but as artefacts that humans can master by virtue of their divinely inspired competences. This radical sense of biblical literalism is epitomised in the Faustian desire to follow the example of God in the Gospel of St John, which explicitly portrays the divine thought processes as the enactment of a drama of one's – originally God's – own scripting, *logos* in Greek. The contemporary human equivalent is the re-engineering and outright manufacture of organisms through biotechnology, which presupposes mastery over the genetic code as the literal grammar of life: humans can create just as God creates. I shall return to this scientifically empowering interpretation of biblical literalism throughout this book, most notably in chapter 7.

Bacon's radical biblical literalism reflects a strand of the Protestant Reformation whose standard-bearer in the Western literary tradition is the devil-contracting Johannes Faustus, a devotee of the Socinian heresy, nowadays known as Unitarianism. Named for the Sozzini family, which included a member called Fausto, these heretics were suspected by 16th-century Catholic authorities of trying to arrogate divine powers, having interpreted humanity's creation in the image and likeness of God in what theologians call 'functional' terms. That is, we were created to become God, at least to complete his work – if not replace him altogether, as in the 18th-century Deist image of *deus absconditus* ('God disappeared'). The modern concept of progress flows from this heresy, which implies our convergence on the ultimate representation of reality – not simply an endless increase to knowledge. By the end of the 17th century, this image of convergence had begun to be clarified with Newton's and Leibniz's independent invention of calculus, a mathematical technique for constructing successive approximations of some anticipated outcome, as in the ancient geometric puzzle of 'squaring the circle'.[28] The relevant theological outcome was something God intended but would not, or perhaps even could not, achieve by himself. Humans are thus needed to correct nature's remaining imperfections and thereby realise its hidden potential.

Credit for making explicit this targeted conception of progress – including its subtle but significant shift in focus from the divine plan as *always already perfect* in sacred time to the plan's *coming to be made perfect* by humans in secular time – goes to Joseph Priestley, the late-18th-century Yorkshire chemist and cleric who was the house philosopher of the high-minded industrialists attached to the Lunar Society, Birmingham's religious nonconformist answer to the Royal Society of London.[29] Before science was organised along disciplinary lines in the 20th century, Priestley was unequivocally credited with the discovery of oxygen. Consider Priestley's standing in 1874, the 100th anniversary of his discovery. Darwin's staunch ally, T.H. Huxley, delivered the main commemorative lecture in the UK. In the US, the centenary provided the pretext for founding the American Chemical Society, which still annually awards the Priestley Medal for the highest achievement in chemistry. Yet today, Priestley himself is normally treated as an also-ran to Antoine Lavoisier, who gave oxygen its name and with whom Priestley corresponded for several years over how to interpret the relevant experiments. Priestley's historical degradation is evident in popular histories of science, where he routinely appears as consumed by 'religious fanaticism'.[30] More likely, Priestley remained a bit too close to the strong Faustian reading of biblical literalism, which harked back to chemistry's roots in alchemy, which in turn, as we shall see, inspired a more aggressive policy towards the creation and management of life than Darwin thought wise.

Alchemy: Intelligent Design's Faustian Underground

Today 'alchemy' is mentioned in the same breath as 'magic' and even 'mystery cults'. This started only once Pope John XXII drove the alchemists underground in 1317, accusing them of infringing on divine patent law (aka natural law) and potentially flooding the market with counterfeit goods – not least fake gold – which they promoted as better than nature's, and therefore God's, original products. Afterwards, alchemists evaded papal censure by cloaking their theories and practices in allegorical language drawn from biblical and mythical sources. Successive generations of alchemists were so successful in this regard that they managed to confuse each other across the centuries, resulting in the promulgation of far-fetched claims that made alchemists frequent targets of derision. However, in the century prior to papal censure, supporters of alchemy – again in England – envisaged a radical understanding of the relationship between science and religion that was more synergistic than segregationist. This, I believe, provides the most useful precedent for ID theory as a scientific research programme.

The 13th century witnessed the first major academic rivalry: Oxford vs. Paris. Both universities flourished in the wave of Latin translations of Aristotle from Arabic, the language in which the classical philosophical and scientific heritage had been systematised, transmitted and enhanced through the Roman Empire's slow disintegration. Influenced by Islam's self-understanding as the culmination of all sacred and secular knowledge, Arab scholars had been the first to conceptualise the synthesis of Judaeo-Christian and Greco-Roman writings as a common knowledge base for humanity.[31] The question that loomed for them, and now their Latin legatees, was how to capitalise on this base. Two generations of Latin scholars – the Dominicans, Albertus Magnus and Thomas Aquinas, in Paris and the Franciscans, Robert Grosseteste and Roger Bacon, at Oxford – offered contrasting visions of what it means to be human and hence how to carry forward in Christendom the legacy forged originally in the name of Islam.[32]

The late anthropologist Mary Douglas drew a distinction that usefully captures the stakes here. According to Douglas, societies organise their world view along two dimensions – *grid* and *group*. What matters more – how members of a society differentiate themselves as individuals, or what distinguishes them collectively from everyone and everything else? In Douglas's jargon, Paris was 'high grid' and Oxford 'high group' – at least when it came to the place of humans in the divine plan.[33]

The Parisians saw humans as very much embedded in the natural world, first among creatures, but all firmly subordinated to God. On this view, there is much scope for scientific inquiry but ultimately it is confined to the means, not the ends, of divine creation, which is forever beyond human comprehension. Where science can no longer go, one must turn to a theologically informed understanding of the Bible, which provides a metaphorical, but never literal, sense of God's intent. In this spirit, the Paris-based Thomas Aquinas was canonised as a saint in 1323 partly to contain the increasing influence of the dissenting view of the Oxford-based John Duns Scotus that we differ from God only by degree, not kind. Duns Scotus seemed to suggest that the Bible empowers each individual to take its message forward, a position that if taken seriously would undermine the legitimacy of the Church's claim to represent a unified Christian community, which of course came to pass two centuries later with the Protestant Reformation.

It perhaps comes as no surprise that Aquinas died on a papal mission to patch up differences with the Eastern Orthodox Church. The instinctive accommodationism of the Parisian view enjoys many political virtues, which help explain Catholicism's remarkable resilience as a unified global Church in the half-millennium since the Protestant secession. In effect, the Parisian view embraced the idea of multiple truths for multiple realities, a secular variant of which appears in the 'separate but equal' attitude that US courts have upheld with regard to scientific and religious education. What prevents this permissiveness from dissolving into a philosophically odious 'relativism' or 'subjectivism' is that each 'level' or 'domain' of reality constitutes a jurisdiction with its own recognised authority. In short, anything goes, but not all at the same place and at the same time. It is easy, then, to see why a self-avowed Catholic like Galileo would prove problematic: he insisted on one truth for all occasions that is, at least in principle, demonstrable to all beings.

In contrast, the Oxonians were inclined to a literal reading of the Bible that emphasised the differences between all of humanity and the rest of nature: we have a spark of the divine that the rest do not. This is why God chose to communicate through a medium – the book – from which only humans can derive meaning because only we normally create such artefacts. As mentioned earlier, the idea of 'spark' resonates down through the history of science. It has suggested fire, light, electricity and, in the modern period, the semi- and superconductivity of energy and information. From this standpoint, our various carbon encasements – the peculiarities of our individual bodies – are sources of resistance. They are perhaps comparable to the resistance that God faced from nature's materiality in the process of creation that prevented us from serving as transparent media for the conveyance, amplification and execution of divine intent. Where we resemble animals the most, we are the most limited. It is in this broadly ontological, rather than a narrowly ethical, sense that the Abrahamic strictures on 'pleasures of the flesh' should be understood as an inspiration in modern science. In this context, 'spirituality' amounts to the prospect of more efficient material renderings of our humanity, starting with asceticism and ending, perhaps, with the uploading of consciousness to computers.

In short, to redress the damage done by original sin, we must reassess what is salient in our personal experience so as to retune the channel through which we communicate with God. For Roger Bacon (not related to Francis), the abstraction of the laboratory, which removes diversions to the display of 'real' connections among things, was a concrete projection of the ascetic imperative to live a stripped-down existence in preparation for divine revelation. Bacon's contemporaries found this vision sufficiently threatening to the social order to require his imprisonment – militant experimentalism as the original evangelism.[34] Newton's great virtue, nearly 500 years later, was to keep this side of his thought largely hidden from public view. Nevertheless, it may be time for ID theorists to follow Bacon's example and go public with this attitude.

To be sure, Roger Bacon's spirit enjoys a secular afterlife in the project of 'artificial intelligence', and more recently 'artificial life', which presume that thinking and even living may transpire more efficiently in media other than their usual carbon-based containers. Indeed, artificial intelligence is popularly characterised as the search for a 'frictionless medium of thought', which information technologists then try to realise in successive generations of computers. A complementary trajectory applies to medical research, which has been increasingly dedicated to developing prosthetic replacements and enhancements in order to extend 'fruitful' human lives beyond their natural span.[35] Devotees of these trends are divided over how humanity is to escape its carbon chains. Will individual self-discipline remove or replace our carbon-based dependencies – a line that has evolved from the ascetic monks of the Middle Ages to the followers of cyber-guru Ray Kurzweil who believe that 'the singularity is near'?[36] Or will it be through collective effort – a 'revolution of the saints', as first the Puritans and then the Marxists have thought?[37] As it turns out, these two attitudes were combined in the person of Joseph Priestley, whose scientific reputation and religious orientation endeared him to the American revolutionaries Benjamin Franklin and Thomas Jefferson.[38]

Priestley to Mendel: A Non-Darwinian Science of Human Entitlement

The US founding fathers were attracted to Priestley's belief that humanity will rise to its God-given potential only once it organises itself from first principles – the social contract as the second creation – rather than submit to the sort of default mode of hereditary succession associated with the British monarchy. This they saw as capitulation to animal-like forms of hierarchical subordination that merely reinforce our distance from God. After all, if we have been created in the image and likeness of God, then we should be able to do better than mindlessly reproduce the patterns of governance observable in lower creatures. Perhaps unsurprisingly, Priestley spent his final years in exile in rural Pennsylvania, as his Birmingham home was torched by fellow Christians who viewed his vocal support of the French Revolution as an implicit act of treason against Britain's Church-sanctioned monarchy. Priestley, who interpreted 'Enlightenment' as a secularised version of divine illumination, is the lost hero in the tale recounted in this book.

As I indicated above, Priestley's diminished status is a result of his failure, vis-à-vis his French correspondent Lavoisier, to interpret their discovery of oxygen in a way that would be conducive to the development of chemistry as a scientific discipline based on atomic principles. In contrast, Priestley saw his discovery, which he called 'dephlogisticated air', as a technology – an artificial improvement on nature, in keeping with humanity's biblical mission to do what God could, or would, not do by himself. In the spirit of the original alchemists who sought ways to perfect the precious metals, Priestley believed that he had purged the air of its more volatile and noxious tendencies ('phlogiston'), enabling it to bear only unambiguously life-giving properties. Indeed, soda water, that popular elixir of the modern era, was the commercially viable direction that Priestley suggested for his discovery.

However, this does not quite do justice to Priestley's vision of nature's intelligent design, because the distinction between science and technology was never as clearly drawn in his mind as in Lavoisier's – or ours today. Again like the alchemists, for Priestley technological innovation was the royal road to fundamental scientific discovery, not merely the *post hoc* application of scientific principles. It provided the most powerful – that is, most persuasive *and* most efficacious – empirical demonstration that we had communicated with the divine intellect. It would show, as Faust would have it, that humans had mastered the tools of God's trade, largely by learning from God's own handiwork (aka nature), so as then to perfect it. Thus, a technical improvement in, say, the life-giving properties of water would reveal the full power of God's tools in a way that God himself, for whatever reason, had not.

This is in striking contrast with Lavoisier's more orthodox view that experiments merely reveal for human comprehension what had always been present in the constitution of nature. For Lavoisier, the laboratory was not a factory for the manufacture of a better reality but an especially sophisticated observatory for the reality that already exists. The difference between Priestley's radical Protestantism and Lavoisier's conventional Catholicism is transparently in play here. But equally played out is the medieval difference between Oxford and Paris. However, it was Priestley's more theologically inflected scientific attitude that led to the pervasive modern view that results from a properly certified laboratory are entitled to remake the world in its image as a secular crucible of creation.

It is instructive that Friedrich Engels regarded Priestley – more than La Mettrie, Helvetius, Holbach and other self-declared Enlightenment materialists – as the precursor for the dialectical view of nature championed by Hegel, which Engels himself then incorporated into

Marxism. 'Dialectical' here implies the human drive (or will) to refine the crudity of nature, an exercise in separating wheat from chaff, the result of which are products and processes better suited for extending human dominion. By rendering human invention a literal extension of divine creation, Priestley removed any lingering philosophical doubts about the ability of people to take personal and collective responsibility for their fate. According to Engels, Hegel and Marx removed Priestley's lingering theological commitments without undermining his basic message. In this context, Engels expressed some qualified sympathy for 'energetics'; that is, a thermodynamically inspired world view that saw science as the distinctly uphill human struggle to do the most with the least, in standing defiance of entropy. In the 20th century, this rather daunting metaphysical task was gradually brought down to earth with increasingly consumerist ideas of science as making life more 'convenient' for people, typically conceived as individuals.[39]

For all his celebrated polymathy, Priestley is not normally seen as having contributed to biology. Yet here too he exerted a profound, albeit indirect, influence. In the decade prior to his discovery of oxygen, Priestley taught at Warrington Academy near Manchester, the largest school for religious dissenters in Britain. During his tenure, modern history and the natural sciences were for the first time introduced as regular parts of the curriculum. Among those from across Europe who matriculated with Priestley at Warrington was the future Leopold III of the Principality of Anhalt-Dessau. The Prince subsequently established a similarly enlightened academy in Germany, where one Christian Carl André flourished as a prominent agricultural scientist who then went on to organise the Moravian Academy of Sciences, which inspired the breeding experiments of Gregor Mendel, the Augustinian monk who founded modern genetics.[40]

The Priestley–Mendel connection is secured by a shared intellectual sensibility. The Moravian capital of Brno fancied itself the Hapsburg Manchester, a city the Moravians saw through Priestley's eyes as the crucible for realising humanity's biblical mission of creating a heaven on Earth. Just as Priestley had treated oxygen not as an element in its own right but as a technique or artefact to improve water for human consumption, his German-speaking followers saw plant and animal breeding in similar terms. Their term for hereditary factors, what we now call genes, was *Vererbungsfähigkeit* – that is, an inherited capacity to bring about some humanly relevant end, such as purer wool in sheep.

Mendel shared this interest in the rationalisation of nature. It accorded well with the bold Augustinian reading of Genesis, in which the impediments posed by unruly matter force God to execute the creation in stages – over six days. However, Mendel went beyond his precursors in the intensity of his Linnaean ambition to arrive at the original grammar – that is, the elements and rules of combination – that God used to generate the diversity of nature, the full potential of which still has been only partially realised. In other words, Mendel effectively turned Darwin's vision on its head: artificial selection, rather than being a poor version of natural selection, would complete the process begun but left incomplete by natural selection. But unlike Linnaeus himself, as well as his own contemporaries, Mendel conceptualised this problem in thoroughly mathematical terms, which laid the foundations for genetics as a statistical science.[41]

But no sooner had Mendel been posthumously celebrated as the founder of genetics in the early 20th century, than doubts were raised that he had massaged the data from his experiments and misrepresented the claims of other researchers, overestimating the long-term viability of hybrids as new species.[42] Indeed, suspicions had hounded Mendel in his own day, not least because he was a religiously inspired amateur. And perhaps Mendel's theological zeal had unleashed an unbridled optimism alien to harder-nosed naturalists, including Darwinists, who believed that all life forms, including hybrids, naturally regressed to the mean of their ancestors' traits over successive generations. In retrospect, Mendel's reception may provide one of the clearest cases of naturalistic prejudice obscuring a scientifically fruitful supernaturalism. I refer here to the correspondence between Mendel and Europe's leading botanist, Carl van Nägeli, who advised Mendel to test his hypothetical laws of heredity on hawkweed, a plant notoriously difficult to hybridise.[43] Mendel had supposed that it was possible to escape from nature's own version of original sin by careful breeding that elicited its hidden good. But after failing to reproduce his pea findings on hawkweed, Mendel turned away from research to administration. Nevertheless, prior to his discouragement, Mendel had clearly recognised the difference between the traits that a given generation actually expresses on its own body, and those that it is capable of transmitting to subsequent generations to be expressed under the right conditions – that is, 'phenotype' and 'genotype', in modern biological jargon.

Mendel's attitude carried over to the population geneticists who remained active in animal breeding, most notably Sewall Wright, another Unitarian, who was perhaps the biggest influence on the original presentation of the Neo-Darwinian synthesis, Theodosius Dobzshansky's *Genetics and the Origin of Species* (1937). While working for the US Department of Agriculture in the 1920s, Wright tried to expedite the evolution of cattle by the regular segregation of herds into mutually exclusive mating communities, which he believed would strengthen natural selection's hand in counteracting the effects of random genetic drift that might otherwise undermine whatever 'progressive' traits had emerged in the cattle.[44] That such a policy might succeed in the long term – that humans might bend the course of natural selection to satisfy their desires – was alien to Darwin's entire line of thought. Admittedly, what we would now describe as Darwin's 'precautionary' attitude absolves him from any complicity in the atrocities committed in the name of eugenics in the 20th century. But it also raises questions about just how constructive a force Darwin would have been, had he lived long enough to witness developments in applications of biology associated with medicine and agriculture.

CHAPTER 3

Is There a Middle Ground Between Creation and Evolution?

The Manufacture of the Science–Religion Conflict

The idea that there always has been – and always must be – an ideological struggle between science and religion was manufactured in the final quarter of the 19th century, specifically to provide a world-historic pretext for Darwin's challenge to religious control over all levels of education, even in nominally secular countries. After all, if Darwin had demonstrated that the detail of nature failed to offer evidence of ID, it was no longer clear what sort of cognitive authority theology could legitimately exert over what is taught in schools and universities.

Darwin claimed that whatever sense we make of nature beyond the need to survive is really just a matter of luck, not representative of the rational structure of reality that, in turn, may reflect some special relationship with the deity responsible for it. In that backhanded way,

Darwin unwittingly exposed just how much science had been always based on a faith in human reason's ability to transcend its immediate circumstances. Even the decidedly anti-clerical John Stuart Mill realised in his review of *On the Origin of Species* that Darwin literally denied the intelligibility of nature.[1] Thus, Mill joined his lifelong enemy William Whewell in fearing that beyond depriving theology of its traditional subject matter, Darwin would also breed a scepticism towards science itself that would license irrationalism and possibly even nihilism. By the time of Darwin's death, such fears had begun to be realised in the writings of Friedrich Nietzsche. And as I have suggested, the problem remains today in the poverty of evolutionary explanations of science itself. In short, how do Darwinists ensure that they do not throw out the scientific baby with the theological bath water?

This question was handled in Darwin's day by rewriting intellectual history to show not only that reality could be rendered rational without religion, but also that the triumph of reason required the outright rejection of religion. The two main histories published in Darwin's wake, *History of the Conflict between Religion and Science* (1874) by John William Draper and *History of the Warfare of Science with Theology in Christendom* (1895) by Andrew Dickson White, were written by, respectively, the first president of the American Chemical Society and the first president of Cornell University. Both saw the conflict between science and religion in terms of prior claims of religious authority halting the course of empirical scientific inquiry. Draper interpreted the problem in Catholic terms – namely, the obstacle posed by papal infallibility – while White tackled the more Protestant-style dogmatism associated with biblical literalism. A big part of Draper's and White's achievement was to leave succeeding generations with the impression that the advance of science had always been stymied by religion when, on balance, the opposite has been the case. Even Darwin suffered no religious censorship in his lifetime.

However, Draper and White were helped by the ongoing publication of the original documents and commentary surrounding the Papal Inquisition of Galileo. Although the events had taken place around 250 years earlier, the archives of the Holy Office were opened only in 1849. Soon thereafter, a steady stream of secular and religious versions of Galileo's trials were published, each drawing selectively on the full range of materials made available. Whatever the original intentions, it left the general impression that the Church had acted harshly toward Galileo and, more importantly, that this had always been characteristic of the Church's attitude toward science. That the Pope fully exonerated Galileo only in 1992 has done little to dispel that impression.[2] Indeed, the Catholic Church's protracted suppression of the Galileo trial transcripts, combined with its grudging admission of error after the fact, must count as the longest-running public relations disaster in history. It continues to poison the tripartite relationship between science, religion and society in the West, even after four centuries.

If this judgement seems harsh, recall that Galileo's Catholic opponents upheld a conciliatory doctrine of multiple truths, rather like the judgement delivered in *Kitzmiller*, designed to maintain the social cohesion of Christendom in the face of Protestant reformers. To them, Galileo appeared needlessly belligerent. That original motivation should have allowed the Church a more generous response to Galileo *ex post facto*. In any case, as we shall see below, the doctrine of multiple truths remains alive and well among today's so-called theistic evolutionists, who think of themselves as mediating the excesses of, on the one hand, metaphysical naturalists like Daniel Dennett and Peter Singer who happily extrapolate from 'is' to 'ought' across species, and, on the other, supernaturally informed ID theorists for whom science fulfils a divine plan that positions humans as its executors and perhaps completers.

Nevertheless, the prevailing image of the religiously persecuted Galileo explains why historians like Draper and White could so easily cast Darwin in the role of 'heretic', despite his never having had to suffer anything like Galileo's fate. Consequently, the science–religion conflict is often treated as continuous with the history of heretics, who among themselves have agreed on little more than the idea that sincere monotheists need not be united on doctrinal matters. Indeed, they are best seen as intellectual libertarians who would risk their personal salvation on the idea that God might tolerate difference, if not outright concur with their point of view. Heretics have always posed a danger, *especially* to those who saw the hand of God in the vicissitudes of nature. Thus, for much of religious history, publicly visible heretics would be scapegoated for calamities that befell a community – as dramatised in *An Enemy of the People* – in the hope of restoring the natural-cum-social order.[3] In the end, heretical dissent tends to be vindicated. But timing is everything in intellectual history, and the only person likely to benefit from the rehabilitation of a long-forgotten heretic is the historian whose reputation is made by it.

There are exceptions to this rule, of course. They come from people who read history mindful of a universal standard of justice but unburdened by the historian's own guild concerns. They take the intensity and duration of what has been deemed heretical as symptomatic of the depth of the truth being suppressed. Though not infallible, this judgement is not unreasonable. Insofar as 'truth' meaningfully characterises the direction of organised inquiry, its aspiration is always universal in a dual sense: truth is not only universally applicable, but also universally accessible. In other words, truth binds all who may know to all that may be known. The very presence of dissent means that this dual sense of universality has yet to be achieved; the dissenter effectively assumes that something is not known by someone and, crucially, that it is an open question – a matter of discretionary judgement – where the fault lies. It may lie equally with an elusive reality, an exclusionary elite, and/or the ignorant masses. The binding character of truth – of everyone to everything – is often obscured (or perhaps not taken sufficiently literally) because science tends to objectify acts that are treated subjectively by religion. Thus, whereas in religion one takes decisions that others may deem heretical, in science heresy becomes error based on a failure to weigh evidence properly. But these are alternative descriptions of the same act. They arouse such great passions only because, by virtue of the monotheistic roots of modern science, we are socially contracted to inhabit a common reality.[4]

To be sure, *Origin of Species* was roundly criticised by religiously inspired scientists, philosophers and commentators who questioned Darwin's hazy views on heredity. But this criticism did little to halt the spread of Darwin's influence. If anything, the controversy surrounding his work aroused curiosity and suggested further research leads, fuelled by those whose causes might benefit from what might be learned. In the 1940s, the US historian Richard Hoftstadter summarised this tendency as 'Social Darwinism', a public philosophy avowed by many industrialists and social scientists in the late 19th and early 20th centuries that attempted to return Darwinism to his inspiration for the concept of natural selection, namely, Thomas Malthus' pessimistic theory of population, according to which conflict over resources is an irremediable aspect of the human condition. Whereas Herbert Spencer, the *Economist* editor who popularised the slogan 'survival of the fittest', could be easily dismissed as having generated an entire metaphysics from some recurrent features of human history in conformity with Malthus, the same could not be said of Darwin.

Darwin elaborated for many kinds of plants and animals the specific terms under which the struggle for survival occurred in nature. Natural selection did not appear as in Spencer's promiscuously proliferated analogies – the precursor of Daniel Dennett's crude and bloated 'pan-selectionism' – but as a subtly emergent feature of nature that merited further empirical study. It was in this spirit that John D. Rockefeller invested millions of dollars in social and biological research; in the hope that humanity's adaptive capacity might be at least marginally improved over the odds on offer in nature. As it turns out, by 1932 the administration of Rockefeller's vision would fall to an ID proponent, the mathematician Warren Weaver, a Wisconsin progressive of Unitarian leanings who had a much stronger faith in humanity's capacity to transcend its material inheritance. Weaver, a pioneer in information theory, was the person most responsible for seeding the revolution in molecular biology on both sides of the Atlantic. We shall encounter him again in chapter 6.[5]

Moreover, in Darwin's day, the role of religion in public life was under intense scrutiny as democratic movements, typically in civic republican guise, targeted established churches as barriers to full participation in civil society in two senses. On the one hand, churches encouraged submission to authority, contrary to the independent thought required of the voting citizen; on the other, clerical authority, epitomised in a meddlesome Pope, might command citizens to act in ways that undermined the national interest. These fears placed religion on the defensive. Not surprisingly, then, contrary to the secular rhetoric of the time, the 'science vs. religion' conflict was more about science dislodging religion than religion impeding science.

Perhaps the most explicit and systematic reorganisation of the sciences in anti-clerical terms occurred in France during the Third Republic, in the final quarter of the 19th century. It was here that Darwin, whose influence on French science has been otherwise relatively muted, was mobilised to wrench the emerging sciences of psychology, sociology and anthropology from theological oversight.[6] In all of these science–religion confrontations, the religious side was guilty mainly of complacency in the face of the assault. For example, in the famous 1860 Oxford Union debate between T.H. Huxley and Bishop Samuel

Wilberforce, Wilberforce misjudged just how much the tide had turned against authoritative appeals to religion in public affairs. Nevertheless, he may have been correct that the point of pursuing pure science would become radically unclear without appeal to a divinely inspired human reason. Indeed, Huxley may have secretly agreed, given his own reluctance to have evolution taught in schools.

A subtle sign of the changing times was the formal establishment of the doctrine of papal infallibility, which happened only in 1870. Once the newly unified Italian nation stripped the Catholic Church of its imperial pretensions by confining it to a precinct in Rome, the Pope adapted by limiting his authority to matters specific to religious doctrine, but where his word would then be taken as final. The motivation for this position is comparable to that which informs the pronouncements periodically issued by the Royal Society or the US National Academy of Sciences: they are officially cast to protect the integrity of science rather than to dictate public policy. But the strategy in both the sacred and secular cases is fairly transparent. If you lack the material and ideological resources to contain dissent on a broad front, you retreat to a narrower but still significant domain within which you might reasonably silence opponents.

Theistic Evolution and the Teilhardian Challenge

To acknowledge the manufactured character of the science–religion conflict is not to deny the reality of its effects. These cannot be wished away by the intellectual mirage known as 'theistic evolution' that points to an illusory but inviting middle ground that would segregate science and religion as if they were races. If a mark of pseudoscience is a refusal to subject one's most deeply held beliefs to rigorous empirical tests, the mark of pseudo-religion is a refusal to hold one's empirically grounded beliefs to a higher cognitive standard that would unify and provide meaning and direction to them, perhaps in ways that force one away from the default stances of the scientific establishment. In this respect, ID theory is much less pseudoscience than theistic evolution is pseudo-religion – religion with all of the heart but none of the brain engaged.

Theistic evolution receives initial credibility from Stephen Jay Gould, himself agnostic, who put the case as follows:

> Science can work only with naturalistic explanations; it can neither affirm nor deny other types of actors (like God) in other spheres (the moral realm, for example). Forget philosophy for a moment; the simple empirics of the past hundred years should suffice. [Gould then tallies evolutionists from Darwin onward who did and did not believe in God.] … Either half my colleagues are enormously stupid, or else the science of Darwinism is fully compatible with conventional religious beliefs – and equally compatible with atheism.[7]

But these are not the only two options. A third option is that religiously inspired evolutionary scientists minimise the significance of the distinctly Darwinian process of undirected natural selection.

An instructive case in point mentioned by Gould is Theodosius Dobzhansky, author of the book that arguably consolidated the Neo-Darwinian synthesis: *Genetics and the Origin of Species* (1937). A devout Eastern Orthodox Christian, he is also credited with the favourite one-liner of evolution's defenders: 'Nothing in biology makes sense except in light of evolution.' The claim appeared as the title of an address to biology teachers in 1973. However, as the conclusion reveals, the lecture was not about defending Darwin:

> One of the great thinkers of our age, Pierre Teilhard de Chardin, wrote the following: 'Is evolution a theory, a system, or a hypothesis? It is much more. It is a general postulate to which all theories, all hypotheses, all systems much henceforward bow and which they must satisfy in order to be thinkable and true. Evolution is a light which illuminates all facts, a trajectory which all lines of thought must follow. This is what evolution is.' Of course, some scientists, as well as some philosophers and theologians, disagree with some parts of Teilhard's teachings; acceptance of his worldview falls short of universal. But there is no doubt at all that Teilhard was a truly and deeply religious man and that Christianity was the cornerstone of his worldview. Moreover, in his worldview science and faith were not segregated in watertight compartments, as they are for so many people. They were harmoniously-fitting parts of his world view. Teilhard was a creationist, but one who understood that the Creation is realised in this world by means of evolution.[8]

Pierre Teilhard de Chardin was a heretical Jesuit palaeontologist who helped discover the Peking Man fossils in 1929. Dobzhansky devoted the final chapter of his most philosophical book, *The Biology of Ultimate Concern*, to a critique of 'The Teilhardian Synthesis'.[9] The critique repays further study today as an attempt to accommodate Teilhard's 'creative evolutionism' to then-recent discoveries in molecular biology, especially DNA. Of particular interest is Dobzhansky's development of Teilhard's concept of *tâtonnement* ('groping') as evolution's sense of purposefulness. This is the idea that genetic variation deployed in a differentiated environment over time tends to converge towards certain forms, the exact ultimate identities of which cannot be known in advance, but which nevertheless provide the best explanation for the overall pattern that evolutionary history displays.

This line of inquiry had already been advanced in Darwin's day by one of T.H. Huxley's own students, St George Mivart. He was a devout Catholic who argued that the seemingly endless differentiation of species masked a deeper sense of long-term morphological convergence towards which all life forms are striving, which explained why the range of fundamental blueprints for life appears to have narrowed over time.[10] Regardless of its historical origins, which Mivart granted might be multiple and even specially created, life must proceed along a generally prescribed course – allowing for resistance from its diverse environments – to perfect certain universal life-functions. This view effectively turned Darwin on his head.[11]

Teilhard went further than Mivart in inferring evolutionary convergence. He regarded the increasing dominance of humanity on the planet – even at the expense of other species – as the vanguard of this singular evolutionary process that will eventually convert the Earth into a 'hominised substance', a dream also shared by the most visionary Enlightenment *philosophe*, Marquis de Condorcet, whose optimistic vision of population growth was specifically targeted by Thomas Malthus.[12]

Teilhard so challenged *both* biblical creationists and Darwinian evolutionists that he dashed their most promising hope for harmony. Both sides have celebrated – and more importantly wished to preserve – the diversity of nature, albeit on somewhat different grounds: on the one hand, biblical creationists regard each species as having been specially created by God with its own essence and purpose. On the other hand, Darwinian evolutionists believe that these apparently essential differences are really emergent features of populations that diverged at some point in the remote past, and since that time have successfully reproduced among themselves in geographically segregated regions that together constitute a natural ecology. Prophet of sociobiology and biodiversity E.O. Wilson has recently tried to capitalise on this point of agreement to organise sacred and secular forces on behalf of the world's ecology.[13]

As opposed to all this, Teilhard took his cue from the originator of modern evolutionary theory, Jean-Baptiste Lamarck, who denied the significance of species differences altogether. For Lamarck, life is simply intelligence striving to transcend its material limits, a thesis notoriously exploited as 'the inheritance of acquired traits'. Thus, nature comes to be unified through the interaction of living things, each trying to learn from the rest, the legacy of which is then transferred to the next generation. While Darwin's first major experimental supporter,

August Weismann, is normally credited with having put paid to this idea by demonstrating that somatic changes do not leave a trace on an organism's genetic material, Lamarck has been revived every other generation, nowadays in the idea that 'horizontal gene transfer', which probably occurred in the very remote past, occurs spontaneously nowadays in the spread of certain diseases and is likely to increase in the future through explicit biotech interventions.[14]

Teilhard exploited the theological potential of Lamarck's biological vision, according to which organisms are best seen as rough drafts of a plan whereby God is fully embodied in material reality, in the course of which nature is brought to self-consciousness. Indeed, Lamarck coined the term 'biology' to name just this project, taking rather more seriously God's creative logos in the New Testament than today's biologists are inclined to do. Like Lamarck, Teilhard equivocated between the theistic and outright humanistic interpretations of this goal. In their telling, evolution may be read as resulting in either our finally bringing nature into conformity with God's idea or God himself finally coming to full realisation through our domination of nature. Is 'the meaning of life' something discovered or invented? What continues to make Teilhard's creative evolution so controversial is its refusal to recognise a difference between the two answers.

Dobzhansky was not the only founder of the Neo-Darwinian synthesis to be impressed with Teilhard. Teilhard tapped into what Julian Huxley, himself a secular humanist, dubbed the 'transhumanist' dimension. This dimension, whereby science would confer on us divine powers, is traceable to the Eastern Orthodox doctrine of theosis, the source of the common word 'apotheosis', a state that St Athanasius, a 4th-century Pope, characterised (with reference to Jesus) as 'God became human so humans could become gods'. Nowadays 'transhumanism' refers to various attempts to 'enhance' evolution, if not to move altogether from carbon to silicon embodiment, as in the uploading of human consciousness into computers in the cause of perpetual existence. Except in the mind of the great prophet of 'spiritual machines', Ray Kurzweil, this prospect currently belongs to science fiction. However, it takes seriously the theological proposition that if there is a spiritual character to our being, then it should be possible to embody it in multiple ways without major corruption. Teilhard's own version of transhumanism consisted of inferring that successive biological species are progressively better drafts of what we shall ideally become.

Unfortunately Teilhard had a rather gnomic mode of expression, which has often been dismissed as 'mystical'. Nevertheless, he clearly realised that taking seriously the idea that God works through evolution requires a rethinking of not only religion but also science itself – by which he meant both the best explanation and the best projection of the empirical character of evolution that has culminated in what Teilhard called the 'phenomenon of man'. His main philosophical works were banned from publication by the Roman Catholic Church until after his death in 1955 precisely because he refused to uphold what is now the signature thesis of today's theistic evolutionists; namely, a neat division of labour between science and religion in terms of 'how' and 'why' questions about the nature of life.

Theistic Evolution and the Darwinian Challenge

Teilhard's bold synthesis may be instructively contrasted with the dubious intellectual ground staked out by contemporary theistic evolutionists. Our first witness is the poster boy of science–religion segregationists, Francis Collins, the born-again Christian who led the US National Institutes of Health's drive to map the human genome. His recent best-seller, *The Language of God*, recounts how his bohemian upbringing resulted in a spiritual emptiness that only came to be filled upon reading C.S. Lewis' *Mere Christianity* as a graduate student in biochemistry.[15] This small fact is telling. Lewis, a colleague of J.R.R. Tolkien at Oxford, is often recommended to open-minded people in order to ease them into the Christian faith. Lewis' rhetorical gifts lay in presenting Christianity as demanding little more than what readers are already presumed to believe, but promising much in return if they just take the extra step to accept Jesus as their personal saviour. In a sense, Lewis merely updated the spiritual eclecticism of medieval literature – his scholarly expertise – that had been used to ease the passage of the British Isles' pagan natives into Christendom. It is a low-cost approach to religion that reappears in Collins' eagerness to play down any important conflicts between God and Darwin. He wants to square things for scientists who don't want ID on their doorstep but who also don't want to have to examine their own beliefs too closely.

Collins, long active in the American Scientific Affiliation, perhaps the world's largest Christian-oriented scientific society, explicitly recommends reading the Bible literally when it comes to morals and ultimate meaning (i.e. in answer to 'why' questions) but not to statements about the natural world (i.e. in answer to 'how' questions), in which case he advises deference to the latest advances in the physical and biological sciences. To be sure, such schizoid hermeneutics has long been second nature to Christians who routinely profess to do science during the week and religion at the weekend. Nevertheless, it raises the question of how the probative value of a supposedly sacred text can deteriorate at such radically different rates over time, such that the Bible's answers to 'why' questions remain intact after two or three millennia, while its answers to 'how' questions hold no validity whatsoever today. Of course, it is only to be expected that new evidence will bear differently on different knowledge claims. But why, in the case of the Bible, should this difference correspond so neatly to the conceptual distinction between 'how' and 'why' questions?

Suspicions are fuelled by Neo-Darwinist counter-claims that answers to *both* 'how' and 'why' questions need to be revised in light of scientific evidence that has become available only in the recent past. Most notably, evolutionary psychologists have begun to cast doubts on the selective advantage of adopting a pro-human universalist standpoint to morality, given our persistent failure to live up to the strictures of either the Judaeo-Christian ethic or its secular counterpart, Kant's categorical imperative, no matter how much pressure of various sorts is placed on us. Our spontaneous sense of benevolence extends to immediate kin and acquaintances, our pets and maybe even our computers, but not to the millions of starving and tortured people in Africa – and this is not due to want of publicity.[16] Africans, it seems, do not persuasively figure in our sense of a sustainable genetic ecology. Perhaps, then, the evolutionary psychologists argue, we should reorient our ethical horizons so that we are not doomed to eternal self-flagellation. In this spirit, Peter Singer has called for the political left to exchange Marx for Darwin, and Steven Pinker has called more generally for an end to utopian social policies based on innate human equality. To be sure, these calls continue to meet considerable resistance, as DNA Nobel Prize-winner James Watson recently discovered when he was hounded out of the United Kingdom for telling a journalist that the West is wasting its money on development aid to Africa.[17]

Already in *The Selfish Gene*, 30 years ago, Richard Dawkins could be found chastising the founder of sociobiology, E.O. Wilson, for interpreting the Neo-Darwinist theory of 'reciprocal altruism' to mean that individuals are genetically predisposed to care for members of their own 'kin' in the strong sense of group selection. Such a reading had led critics to brand Wilson a closet racist. But strictly speaking, Dawkins observed, the theory implies that individuals of all species are pre-programmed life insurance underwriters who invest their energies in others less in terms of their species-relatedness than their likelihood to advance their shared genes.[18] Given the 95+ per cent of genetic overlap between humans and most animals, this can make for some very interesting albeit unconscious calculations, resulting in what can be only called a 'brave new moral economy', in which we might conclude that some animals are better investment risks than some humans: that little bit more genetic closeness may not be worth the cost of sustaining 'our own' at the expense of 'them'.[19]

It is no accident that Peter Singer has been promoting the moral status of animals, while campaigning for the easing of restrictions on abortion and euthanasia in humans. Revered and loathed in equal measure, Singer is distinguished by the forthrightness with which he strives to keep his answers to 'how' and 'why' questions in Neo-Darwinian synchrony. If classical liberalism aimed for the most freedom that each human can enjoy without denying similar freedom to others, Neo-Darwinian liberalism aims for the most freedom that each *species* can enjoy without denying similar freedom to other species. This latter, so to speak, 'meta-liberalism' may require the curtailment of human reproductive capacities, if not actual lives, analogous to liberalism's constraints on human action – in both cases in the name of maximising a jointly realisable sense of freedom, but now in a space better characterised as an 'ecology' than a 'society'. Nazi Germany was the first regime to take meta-liberalism seriously, albeit crudely and partly to mask other, less scientifically justified motives.

Theistic evolutionists routinely dismiss this entire line of thought, which I have dubbed the 'casualisation of the human condition', whereby the expanding moral horizons of liberalism slide into an anti-human meta-liberalism.[20] They simply take what Collins calls 'the existence of the moral law and the universal longing for God' as a feature of human nature that is entrenched enough to be self-validating.[21] But is their dismissal anything more than an arbitrary theological intervention? If humans are indeed, as the Darwinists say, just one among many species, susceptible to the same general tendencies that can be studied in the same general terms, then findings derived from methods deemed appropriate to animals should apply to us as well. Collins' own comprehensive but exclusive training in the hard sciences may explain why he believes in a God who communicates straightforwardly through the natural sciences but appears less willing to cooperate with the social sciences, including such biologically inflected fields as sociobiology and evolutionary psychology. Instead Collins finds intuition, anecdote, theology and sheer faith to be more reliable sources of evidence. Why God should have chosen not to rely on the usual standards of scientific rigour in these anthropocentric matters remains a mystery.

At the same time, Collins does not wish to privilege his own commitment to Christianity among the world's religions. Ironically, like many atheists, Collins gives all religions equal treatment – only he treats them with respect. This 'broad church' approach to religion results in a philosophy sufficiently devoid of controversy, if not content, to be 'espoused by many Hindus, Muslims, Jews and Christians, including Pope John Paul II'.[22] Here Collins could have done with a course in sociology or anthropology. The very idea that ways of being as divergent as, say, Hinduism and Islam should be lumped together as 'religions' is a remnant of 19th-century attempts to understand how complex social relations managed to survive over long stretches of space and time without the modern nation-state – or their imperial extensions, in the case of India, where Hindus and Muslims were both held in check by the British. As science came to be identified with social and economic progress, itself an overall policy mission of the nation-state, 'religion' unsurprisingly became the generic name for everything that blocked this trajectory of 'modernisation'. This is an important context for understanding the hostility to organised religion in public life found in such liberal thinkers as T.H. Huxley in the UK and John Dewey in the US.

But beyond this common negative definition of 'religion', particular religions have related to science in particular ways. Here Collins fails to deal squarely with Christianity's historical uniqueness as the source of *both* greatest inspiration and strongest resistance to the progress of science. Some Christians have radicalised humanity's biblical entitlement beyond the mere understanding of nature to directing its course and realising its outcome. Other Christians have balked at theories like Darwinism, however well-supported by the evidence, which undermine the basis for this privilege. In this respect, much of the alleged war between science and religion has been fought over the soul of Christianity. Whatever their other faults, ID theorists grasp this point much better than Collins. When an ID theorist thinks of a Darwinist, she doesn't imagine a sanguine science–religion segregationist like Collins who sees evolutionary theory as a boon to preventive and regenerative medicine. Rather, she imagines an animal rights protester who follows Peter Singer in wondering, on good Darwinian but anti-Christian grounds, why human comfort should take priority over animal suffering.

At the very least, Daniel Dennett, Steven Pinker and other purveyors of the pan-naturalistic 'third culture' featured on the Edge website (www.edge.org) are owed an explanation for why someone like Collins should open the door to Darwinist explanations in the first place, only then to shut it at the threshold of the human condition. Here theistic evolutionists cannot be allowed the easy philosophical option of invoking the 'naturalistic fallacy'; that is, the proscription against deriving what we should do from what we normally do. After all, the likes of Collins generally seem quite happy to commit the naturalistic fallacy when it comes to the moral life of *animals*. In contrast, philosophers, including distinguished secular ones, who believe on principled grounds that we cannot derive 'ought' from 'is', tend to regard Darwin's theory of evolution very much as defenders of ID do, a 'theory' in the weak sense of an extended hypothesis that is only as good as the phenomena it reliably explains and, preferably, predicts.[23] From this standpoint, theistic evolutionists simply want to have their cake and eat it.

Nevertheless, no trial on the relative merits of evolution and creation would be complete nowadays without a theistic evolutionist as an expert witness, preferably one who has taken Holy Orders at some point in his life and/or can profess to regular churchgoing and conservative values. No one exemplifies this pharisaic role better than Francisco Ayala, evolutionary biologist at the University of California at Irvine and former president of the American Association for the Advancement of Science, as well as recipient of the highest scientific honour granted by the US government, the National Medal of Science. Ayala debuted the role of theistic evolutionist as an expert witness in *McLean v. Arkansas* (1982), where he testified to knowing many evolutionists, including himself, who were 'religious'. The credibility of that claim was left unexplored – and still went unexplored when reaffirmed by the star witness at *Kitzmiller*, Kenneth Miller, a fellow Catholic professor of cell biology at the

Ivy League institution, Brown University. I shall examine Miller's alleged refutation of ID in chapter 5, but for now I will focus on Ayala.

In US legal matters concerning 'science versus religion', beliefs about science come under much greater scrutiny than beliefs about religion. Indeed, witnesses are normally given a free pass on their religious beliefs, even though it is the potential influence of such beliefs on the curriculum that provide the pretext for such trials. You can declare yourself a devout Christian under oath without expecting challenges that force you to reconcile your published statements or known behaviours with your avowed religious beliefs. In that case, your testimony has great rhetorical force, as someone whose devotion to God does not render their mind impervious to science. But there is a catch: *you also have to uphold the scientific orthodoxy*. If your views are scientifically heterodox, then the mere mention of the Bible renders your scientific judgement suspect.

One wishes that the US legal system exercised the same diligence in authenticating people's religious beliefs as their scientific beliefs. Ayala, Miller and Collins claim that their scientific inquiries are driven by their faith in God. Yet, as they are the first to admit, the science they do is indistinguishable from those who do not share that faith. One might reasonably wonder: how exactly does their faith influence their science, especially given the enormous import of their religious commitments? Would it not be reasonable to expect their Christian beliefs, assuming they have some cognitive content, to colour the theories they propose and the inferences they draw from the evidence? If not, why should we think that their

Christianity has any impact on their science whatsoever – simply because they say so? Perhaps logical positivists like A.J. Ayer were right, after all, when they dismissed religious utterances as no more than emotional outbursts. In any case, theistic evolutionism appears to be the kind of religion that even Richard Dawkins could love, since it appears to exact no psychic cost from its scientific adherents. Their religious beliefs spin as decorative but cognitively idle wheels. What follows? Not necessarily that theistic evolutionists are liars. But if not, then either their theism must be very weak or it is held in a state of captivity, as if they fear its public expression would invite persecution.

One day, a clever lawyer defending ID will undermine the credibility of a Miller or an Ayala in the courtroom by demonstrating that their scientific judgement has been compromised by their religious beliefs, which predispose them to defer to the reigning orthodoxy, thereby suppressing the scientific imperative to remain critical and open-minded. I single out Miller and Ayala because of their Roman Catholicism, the Christian denomination that makes deference to authority most integral to the expression of faith. Thus, Catholics may be especially inclined to treat the organisation of science like that of the Church, both of which are underwritten by a body of thought mediated by time-honoured institutions that claim a privileged origin and a closely monitored line of descent, through which all authorised doctrinal changes must occur. In science, the work of the founder of a 'paradigm', in Thomas Kuhn's sense, anchors what then counts as a legitimate extension, application and testing of the science in question, just as Jesus' anointment of St Peter as his successor anchors the legitimacy of all subsequent Popes. It follows that, despite a general awareness that the leaders of even the most august institutions are fallible, ordinary believers in either religion or science are not licensed to decide for themselves what to think through personal study of, say, the Bible or nature.

Not surprisingly, whenever the plaintiffs in *Kitzmiller* wanted to raise high dudgeon, they would conjure the spectre of ID wanting to 'change the ground rules of science', as if this were tantamount to renegotiating the Ten Commandments. Much more surprising, and perhaps even shameful, was the silence of STS scholars, even though especially the constructivists among us have made good careers from demonstrating that the ground rules of science are *always* being renegotiated, and that the appearance of long-term continuity is simply the product of successive generations agreeing to contribute to the same historical narrative.

The problem with simply saying that Miller and Ayala are using religion to humble science by making it more religion-like – that is, by reinforcing science's current Neo-Darwinian dogma – is that the Bible provides nothing comparable to Peter's anointment to underwrite such deference to scientific authority. However, theistic evolutionists are assisted in their dogmatic deference, first of all, by the inordinately equivocal conception of 'evolution' that prevails in public discourse, not least when scientists are involved. 'Evolution' thus imperceptibly slides from a hard-line Darwinist notion, firmly grounded in natural selection, to a more neutral sense based on random genetic drift to, finally, a quasi-Lamarckian conception predicated on the long-term convergence of organic forms, much as

Teilhard thought. But beyond that, theistic evolutionists are also helped, ironically perhaps, by an impoverished understanding of design thinking, underpinned by a lack of theological imagination, that dismisses ID on the simple grounds that the various forms of life, as we currently understand them, do not seem especially well designed.

Theistic Evolution's Selective History of Science: Design without a Designer?

All of the above concerns are conveniently epitomised in Ayala's recent communication in *The Proceedings of the National Academy of Sciences*, 'Darwin's Greatest Discovery: Design without a Designer'.[24] The piece opens with Ayala's understanding of the history of science, which is strangely beholden to Sigmund Freud's *Introductory Lectures in Psychoanalysis* (1916–17). Here Freud sets his own discipline as the third – and possibly final – stage in humanity's evolution out of collective narcissism, after the displacements wrought by Copernicus and Darwin. The idea is that Copernicus dislodged the Earth from the centre of the universe, Darwin dislodged humanity from the centre of the Earth, and now Freud was dislodging reason from the centre of humanity. Michel Foucault is probably the thinker who has made the most of Freud's self-serving genealogy by arguing that it charts the gradual disillusionment with humanity's privileged position in nature.[25] While Foucault is typically seen as a leader in the 'postmodern' questioning of the ontological integrity of what is variously called 'the subject', 'the author' or simply 'man', Ayala is perfectly correct in saying that Darwin was already doing this 100 years earlier. In this respect, contemporary versions of postmodernism and naturalism fit together hand-in-glove.[26] After all, to even a modern-day Darwinist, we humans are little more than the products of marginal differences in a common genetic material, and are subject to forces that we can control only marginally better than other creatures composed of much the same material. Foucault and Freud, like Darwin before them, were more or less fatalists: for them, most of the causation relevant to human existence happens, as Marx once said, 'behind the backs of men'.

The most obvious problem with following Freud's account of the modern history of science as a Copernicus-to-Darwin trajectory is that it turns the greatest scientist of all time, Isaac Newton, into a bit player. Ayala himself mentions Newton only once in a sentence crowded with other 17th-century notables. Of course, Newton built on Copernicus but he drew it together with other work – from Galileo, Kepler and Gilbert – and this enabled him to forge a synthesis that provided the first rational demonstration that humans could successfully aspire to something more exalted than an animal existence. (Of course others, notably Descartes, had tried before Newton but failed.) Moreover, this was exactly how Newton was understood until the early 20th century, when his specific scientific claims came to be overturned by the revolutions in relativity and quantum mechanics. Indeed, Kant made a point of arguing in the preface to *Critique of Pure Reason* that, in light of Newton, we needed a *reversal* of the revolution that Copernicus brought about in human thought: Newton provided renewed assurance that humans are capable of mastering *even* a universe not physically centred on them, presumably because our cosmic centrality is *not* tied to our physical natures. Here it is worth recalling that Freud's account of narcissism focused on one's own body. Kant's confidence was clearly grounded in something beyond that locus of endless fascination.

I dwell on Ayala's curious understanding of the history of science, apparently endorsed by the world's premier academy of scientists, because whatever conception we have about the progress science has made over the centuries is based on some sense of science's track record. This implies a series of judgements about who and what succeeded and failed, as well as whether they have bolstered or overturned existing orthodoxies. All of this, in turn, presupposes a standard in terms of which an overall trajectory may be plotted. In this respect, Ayala outright misrepresents the history of science when he says: 'Darwin completed the Copernican Revolution by drawing out for biology the notion of nature as a lawful system of matter in motion that human reason can explain without recourse to supernatural agencies.' Whatever else natural selection might have been for Darwin – or his successors, for that matter – it was *not* a supplement to Newton's laws, *pace* Ayala. To be sure, there have been biologists who have sought such laws, but their lineage is better traced through the physicists and chemists who migrated to biology in the 20th century, ultimately providing the molecular basis of genetics. For them, 'evolution' lacks an exact meaning beyond a simple description of the natural history of life forms on Earth.

For his own part, Darwin urged cosmic humility in the face of irredeemably chance-based processes that he believed we could only describe but never fully explain. Darwin did not anticipate 20th-century developments in genetics and molecular biology: he didn't think they would be possible *precisely because* of the elusiveness of natural selection. In this context, Darwin's failure to have recourse to 'supernatural agencies' amounted to an admission that the workings of nature are radically alien to human comprehension. After all, the supernatural agency with which he and his readers were most familiar, the biblical deity, conferred on humans a special cognitive privilege, by virtue of our having been created in its image and likeness. It had been on this theological basis that nature was deemed sufficiently intelligible to repay sustained scientific pursuit. It is worth recalling that that scourge of all design-based arguments for God's existence, the Scottish Enlightenment philosopher David Hume, was equally sceptical of experimental science's capacity to provide us with knowledge of ultimate things.[27] And it was precisely against Hume's science-stopping scepticism that Kant wrote *Critique of Pure Reason*.

As it turns out, T.H. Huxley recovered Hume's philosophical reputation from obscurity by presenting him as Darwin's long-lost intellectual ancestor in 1879 in a popular book series dedicated to 'English Men of Letters', in terms of which Hume was meant to appear as an historian. Huxley's juxtaposition of Hume and Darwin was astute. Hume believed that the idea of cosmic order, with or without God, is literally a figment of our imaginations. His famous refusal to infer causation from correlation in nature, the bedrock of modern statistical inference, provided a *dis*incentive to probe beneath the surface of nature, an attitude that persisted in Darwin's confused views about heredity. In effect, Hume rejected the pretensions of both sacred and secular knowledge. If he deserves our continued admiration, then it should be for his even-handed metaphysical deflationism rather than for any claims about science's superiority to religion. For Hume, *both* Newtonian science and revealed religion stray woefully beyond what the senses can reasonably deliver. He reinvented for 18th-century audiences the therapeutic attitude associated with the moral philosophy of the ancient atomists, Epicureanism, which Ludwig Wittgenstein subsequently reinvented for our own times. According to this tradition, philosophy provides consolation for the futility of all metaphysical pretensions, including Newton's claims to have decoded the divine plan. Little wonder, then, that Auguste Comte, whose positivist politics aimed to install a universal scientific religion to replace the Roman Catholic Church, took Francis Bacon, not Hume, for his inspiration.

Moreover, Hume's naturalistic path can actually lead to bad science. In this context, the historian and philosopher of science Larry Laudan has usefully distinguished two types of inductive inference. On the one hand, God-intoxicated scientists like Newton engaged in 'aristocratic induction', what in the wake of *Kitzmiller* is nowadays derogated as 'supernatural' inference, whereby conclusions about the workings of hidden entities are drawn from observed patterns of ordinary entities. On the other hand, God-detoxicated naturalists like Darwin followed Hume's 'plebeian induction' from observable to observable without presuming a deeper sense of reality.[28] Unfortunately, this approach led Darwin to wrongly infer an organism's genetic potential from the sum of the traits displayed in its family history – as opposed to Mendel's patently supernaturalist, yet also mathematically and experimentally tractable, approach to hereditary factors on which the modern science of genetics is based.

This point is worth emphasising because, like many latter-day evolutionists, Ayala appears to be under the misapprehension that Darwin's version of evolution somehow clarified, rather than obscured, the nature of life. Perhaps the easiest way to see this point is by looking at how evolution is routinely defined today in the biological sub-discipline that can arguably lay the oldest claim to Newton-style laws: population genetics. The following is a lightly edited version of an account taken from a website that has been endorsed by the US National Science Teachers Association as being suitable for college and high school students.[29]

> Through mathematical modelling based on probability, Hardy and Weinberg concluded in 1908 that gene pool frequencies are inherently stable but that evolution should be expected in all populations virtually all of the time. They resolved this apparent paradox by analysing the net effects of potential evolutionary mechanisms. They went on to develop a simple equation that can be used to discover the probable genotype frequencies in a population and to track their changes from one generation to another. This has become known as the Hardy–Weinberg equilibrium equation. In this equation ($p^2 + 2pq + q^2 = 1$), p is defined as the frequency of the dominant allele and q as the frequency of the recessive allele for a trait controlled by a pair of alleles (A and a). In other words, p equals all of the alleles in individuals who are homozygous dominant (AA) and half of the

> alleles in people who are heterozygous (Aa) for this trait in a population.[30]
>
> Hardy, Weinberg, and the population geneticists who followed them came to understand that evolution will not occur in a population if seven conditions are met:
>
> 1. mutation is not occurring
> 2. natural selection is not occurring
> 3. the population is infinitely large
> 4. all members of the population breed
> 5. all mating is totally random
> 6. everyone produces the same number of offspring
> 7. there is no migration in or out of the population.
>
> These conditions are the absence of the things that can cause evolution. In other words, if no mechanisms of evolution are acting on a population, evolution will not occur – the gene pool frequencies will remain unchanged. However, since it is highly unlikely that any of these seven conditions, let alone all of them, will happen in the real world, evolution is the inevitable result.

Notice that evolution is defined in purely negative terms – it is basically the result of *anything* that causes gene pool frequencies to deviate from the ideal case described by the Hardy–Weinberg principle. Notwithstanding the idle use of the word 'mechanisms' in the definition, evolution – let alone natural selection – is hardly a law of nature but rather defined as a general concept for a variety of exceptions to the laws of genetics.[31] Under the circumstances, it would be difficult to imagine what *wouldn't* count as evidence of evolution. Evolution in this sense is, as Karl Popper would say, unfalsifiable. It would be as if physicists defined 'physical reality' as whatever causes the strict regularities of Newton's laws not to be observed in everyday situations. To be sure, some philosophers of science like Rom Harré and Nancy Cartwright conclude from this line of thought that the laws of physics do indeed 'lie', and that, more generally, science cannot live up to Newton's universalist aspirations. According to them, Aristotle was right, after all, to have insisted on what Cartwright has dubbed a 'dappled world' housing multiple ontologies.[32] A reincarnated Darwin would probably adopt that line as well, which would lead him to treat the Hardy–Weinberg equation as an artifice of the laboratory and the mathematical imagination. The problem with dismissing the equation in this fashion is that those exact settings are increasingly becoming the crucibles of creation, via biotechnology and computer simulations.

Darwinian evolution's capacity for obscuring the nature of life is epitomised by Ayala's subtitle, 'Design without a Designer'. This mysterious phrase presupposes a curious dualism. Not only are there designed things with a clear designer, namely human artefacts, but there are also supposed to be designed things without any designer. The natural theologian William Paley coined the phrase 'design without a designer' in 1800 *as an oxymoron.* Whatever his other failings, when compared with today's evolutionists Paley had a remarkable sense of intellectual parsimony. He treated *all* designed things as what they literally are: *artefacts*. For Paley and all ID theorists after him, biology and technology are two species of the same genus, namely, 'design sciences', the former concerned with divine and the latter with human design. Thus, Paley notoriously likened the idea of nature as divine artifice with a watch found on a heath.

The clarity of Paley's vision of a science of design that ranged equally over the creations of God and humans reflected a sense of the 'univocity of being' that harked back to John Duns Scotus. Keep in mind that, generally speaking, the Old and New Testaments of the Bible present divine creation differently, the former more mechanical, the latter more organic; God as craftsman vs. parent. Yet these are ultimately alternative accounts of the same 'univocal' process, where the *vox* in question is God's. But rather than leaving it as an inscrutable mystery, we may address this situation scientifically. How might one translate between the languages of mechanism and organism to show that they are referring to the same thing – that we are literally *both* an artefact and an offspring? We have seen that Linnaeus already thought along these lines. From a mechanical standpoint, organism implies planned obsolescence (or death) and self-replacement (or reproduction); whereas from an organic standpoint, mechanism implies indefinite durability, the extreme of which would be the ever elusive 'perpetual motion' machine. This interpenetration of perspectives has informed projects that have gone by the names of 'biophysics', 'bionics', 'biomimetics' and, of course, 'biotechnology'.[33]

Why humans should try their hand at creating things that already exist spontaneously in nature – indeed often with the purpose of enhancing, if not replacing, the natural ones – is a mystery unless we have reason to believe that our powers resemble those of whoever originally created the natural things. Of course, we might imagine that such powers are probabilistic in themselves and/or in our access to them, which would help to explain any appearance of trial and error. However, in this context, the element of chance must be sufficiently manageable to encourage continued experimentation in the face of short-term failure. Given this chain of reasoning, it is understandable that ID's leading scientific defender in the UK, Andrew McIntosh, Professor of Thermodynamics at the University of Leeds, was awarded a large research grant from the Engineering and Physical Sciences Research Council for research on 'biomimetics', a field that applies the insights of natural systems to human ones by regarding plants and animals as bases for new processes and products. In McIntosh's own work, the bombardier beetle is treated as a prototype for improving the payload delivery of military aircraft by showing how it might reignite its engines mid-flight.[34]

Needless to say, a biological science with biomimetics at its centre is not the same as one that stresses 'biodiversity', let alone 'biophilia', to recall two terms coined by E.O. Wilson, who detected in his own insect speciality, ants, the evolutionary foundations of *all* social behaviour, from which not even humans could escape. Indeed, Wilson has appropriated Whewell's 'consilience' – that is, a drawing together of different bodies of evidence into a unified world view – to flatter his own project of sociobiology as a worthy successor to Newton's unified physical science.[35] But in practice, Wilson's terms are much closer to Darwin's own species-egalitarian orientation to life, whereby plants and animals are entities worthy of promotion in their own right and not simply as means for extending humanity's control over nature. Insofar as biodiversity plays any role in McIntosh's thought, it is probably as a strategy for maintaining the widest range of organic resources that can be then, depending on the metaphor used, 'mined' or 'farmed' for scientific and economic purposes.[36]

Recent ID theorists, notably William Dembski, have followed Paley in arguing that what is not designed is the product of either chance or necessity – or perhaps both. Indeed, Neo-Darwinists believe that everything results from both, rendering the category of design superfluous. Interestingly, in our Darwinised times, design tends to be distinguished from chance, but Paley thought it had to be distinguished from necessity. This was due to the precedent set by Newton, who had argued for divine intervention because his own laws predicted that like the ultimate mechanical clock, the universe's machinery naturally winds down without a regular energy boost. In that sense, design complemented necessity in a universe where nothing was, strictly speaking, left to chance. Thus, Paley argued that 'there cannot be design without a designer' in the same sense that there cannot be 'order without choice'.[37] By the end of the 19th century Newton's solution had come to be interpreted in thermodynamic terms, with the ordered state of the universe featuring as an improbable outcome of the laws of statistical mechanics. This suggested to ID theorists like James Clerk Maxwell and Ludwig Boltzmann that the universe was designed to be understood by creatures like us, an idea that is nowadays often called the 'anthropic principle'.

Like most evolutionists, Ayala equivocates on whether Darwin believed that design in nature is real or illusory. Yet it is reasonably clear, despite Ayala's claims to the contrary, that Darwin's considered opinion was that design is illusory. Thus, lacking any overall design in nature to explain, he found no need to postulate an intelligent designer. Darwin, not unreasonably, held a deity worthy of his belief to rather higher design standards than what struck him as being on display in the wasteful and directionless character of natural history. We might regard his judgement in one of three ways, the first two of which are mutually exclusive. On the one hand, he might have been a theological simpleton ignorant of theodicy and hence oblivious to how nature's 'evil' aspects (e.g. sudden mass extinctions) might be reconciled with the divine plan as something fully realised only in the long term; on the other, he was a theological sophisticate who began by agreeing with Paley, who had coined the oxymoronic phrase 'design without a designer' in a spirit of dismissal – but then, much to his surprise, found insufficient empirical grounds to support Paley's premise. Darwin's judgement contained elements of both of these opposites. But he would certainly be surprised to learn that nowadays it is common for Ayala and others to interpret Paley's oxymoron in a third way, without the slightest hint of irony. Indeed, were Darwin transported to our times, he would concede, in light of the largely laboratory-based work in genetics and molecular biology that has transpired since his death, that there *is* design in nature and that he had prematurely dismissed that prospect simply on the basis of the nature of life (and death) as he had observed it in field settings.

Ayala doesn't let the real Darwin get in the way of claiming on his behalf that real design does not require a real designer but could result 'simply' from natural selection, which appeals to chance-based processes to determine the differential reproduction patterns in populations of organisms. Before proceeding, let us linger over what 'simpler' might mean in this context, given the omnipresence of this word in defending the superiority of natural selection to God as a scientific explanation. Such appeals to simplicity are popularly called 'Ockham's Razor', after the 14th-century English scholastic philosopher William of Ockham, who argued that nature's economy dictates that we not appeal to more elaborate explanations when simpler ones will do.

This pronouncement is usually interpreted as an important step towards the more hard-headed empirical approach to reality as exemplified by modern science. However, Ockham originally wanted to explain how it is possible that language enables us to have reliable access to the world, especially given that the existence of things is entirely independent of the words we invent to refer to them. (Keep in mind that Ockham was writing four or five centuries before the advent of serious brain science or linguistics.) Most of the explanations on offer at the time postulated a mediating realm of 'ideas' – either somewhere in our heads or in Plato's heaven – that underwrites the link between words and things. Ockham held that such an explanation merely duplicates the problem, since presumably in that case there would need to be exactly as many ideas as there are words. In addition, it fails to capture the *overall* reliability of our language use. It is not simply that some bits of language work and others fail; rather, it all works (at least when referring to things). For Ockham this sense of unity implied a prior cause: God, an explanation infinitely simpler than a shadow world of ideas.

The point of this digression into medieval philosophy is that Ockham's Razor was originally used, as it is in contemporary ID, to demonstrate God's explanatory relevance.[38] The task of explaining natural phenomena involves much more than explaining a series of isolated events in the world, each individually. It also involves explaining how those events manage to hang together to constitute a coherent world order. Modern science takes off only once this latter task is interpreted as going beyond the simple postulation of God, as Ockham did, and probing in more detail – by both rational and empirical means – the nature of divine intelligence as expressed in the capacity for a few esoteric principles to generate the entire diversity of nature. Newton was of course the great champion of this approach. Kant followed him, except that, in the end, he was uncertain whether it served to illuminate God's or simply our own minds. Darwin, by contrast, rejected this entire line of thought by denying its underlying premise – that there *is* a unity to nature above and beyond the particular things that happen in the natural world. Thus, for Darwin 'natural selection' does not name a simple explanatory concept but the absence of just such a concept. Once again, this point is rarely appreciated, because we mistakenly imagine that Darwin's work was gesturing towards what became experimental genetics and molecular biology in the 20th century.

None of this stops Ayala from giving natural selection a most un-Darwinian mechanistic reading as a 'two-step process', which he presents as follows:

> First, hereditary variation arises by mutation; second, selection occurs by which useful variations increase in frequency and those that are less useful and injurious are eliminated over the generations. 'Useful' and 'injurious' are terms used by Darwin in his definition of natural selection. The significant point is that individuals having useful variations 'would have the best chance of surviving and procreating their kind.'[39]

The word 'chance' in the quote from Darwin is meaningless: if the designed character of nature is as purposeless as Darwin's opposition to Paley would suggest, then 'useful' and 'injurious' are terms that can be attributed only with hindsight, based on which members of a population have actually passed on their genes to the next generation. However, Ayala doesn't make this point as clearly as he might, since he still wants to trade on the old creationist built-to-purpose connotation of 'useful', which implies that the usefulness of a trait could be specified in terms of properties of an organism without having to consider its genetic history. Here one appreciates the brutal honesty of Richard Dawkins, who quite rightly says that from a Darwinian standpoint, organisms are simply more or less successful vehicles for the transmission of genes. And if evolution is 'about' anything at all, it is about this – and nothing more. Saying that a population's increase over successive generations reflects the usefulness of its members' traits is no better than saying that it reflects divine approval. In both cases, a *post hoc* honorific is mistaken for a determining cause.

However, Ayala and other evolutionists are rarely as explicit as Dawkins about the profoundly alienating world in which Darwinism leaves us. This is because natural selection is normally portrayed as being not quite as 'random' as hereditary variation. In other words, the traits in a population of organisms that survive one cycle of reproduction in a specific environment usually survive several cycles. This explains how it is possible for a mutation eventually to become a species in its own right. It is therefore easy – albeit sloppy – to argue that the traits with which a given generation is born 'enable' them to survive, implying that the organisms were specifically designed to be that way. In contrast, if selection environments were subject to as much variation as the hereditary material on which they operate, then all sense of design would probably drop out of evolution talk. But in that case, our lives would be so short, nasty and brutish that science itself would be impossible.

But why don't selection environments change more often? Could life survive at all, if the parameters of those environments varied too often and by too much? These questions raise concerns normally associated with the 'anthropic principle' in cosmology, according to which the physical constants and other robust constraints in nature have had to be more or less as they are in order for life – especially of the human variety – to exist at all. Of course, the anthropic principle permits considerable variation, *but always within those constraints*. This sensibility is familiar to those who think of scientific theorising as a form of model-building, as in laboratory experiments and computer simulations that purport to replicate, if not provide a concrete instance of, evolutionary processes. In these cases, one must adopt a design standpoint: *before* one can generate variety in nature, one must lay down the parameters of permissible variation.

Ayala's two-step presentation of evolution, which is quite common in contemporary biology, thus needs to be reversed. Without prior specification of the selection environment, natural history is unintelligible. Ronald Fisher, the statistician most responsible for providing a secure basis for natural selection in probability theory, famously likened its workings to a casino that turns a profit by manipulating the odds for success; hence the predominance of extinctions (aka losers) in nature. But as a Christian eugenicist, Fisher held God to be the ultimately benign casino owner who hoped his specifically human players might discern his 'chance by design' set-up, so as to beat the house at its own game by strategically planning their own reproduction. Unlike its Nazi variant, Christian eugenics would treat natural selection as less a non-negotiable law that science facilitates, than a challenge for humanity to collectively overcome through self-discipline, thereby rendering us worthy stewards of the planet.[40]

In the case of experiments and simulations, human scientists are clearly the relevant intelligent designers who determine the selection environment by setting the parameters for permissible variation. But who or what constrains the workings of *natural* selection? Darwin himself provides little help, since he believed that natural selection radically differs from the 'artificial selection' long practised by plant and animal breeders. Indeed, his scepticism seems to have run so deep as to cast doubt on whether later laboratory-based attempts either to induce new species or correct deficiencies in old ones would have passed muster in his eyes. Consider the following:

> I have called this principle, by which each slight variation, if useful, is preserved, by the term Natural Selection, in order to mark its relation to man's power of selection … We have seen that man by selection can certainly produce great results, and can adapt organic beings to his own uses, through the accumulation of slight but useful variations, given to him by the hand of Nature. But Natural Selection, as we shall hereafter see, is a power incessantly ready for action, and is as immeasurably superior to man's feeble efforts, as the works of Nature are to those of Art.[41]

In fairness to Ayala, it is true that Darwin's attitude here is entirely consistent with a belief in a divine creator – but one whose power exceeds human comprehension and control to such an extent that comparisons with human artifice are futile. This is just another expression of Darwin's denial of the intelligibility of nature. He regarded the successes of artificial selection as local and temporary, not the sort of thing about which one might construct mathematical formulae of potentially universal scope. It was left to Mendel the Augustinian monk to articulate this grander prospect with his three laws of heredity, on the basis of which the Hardy–Weinberg principle was formulated, upon which the modern science of genetics rests. In effect, when Darwin strayed from a biblical creationist line, he ended up underestimating the capacity of biology to restore and even enhance natural life forms through medicine, biotechnology and environmental science.

'Natural selection' was merely a metaphor for Darwin, not a prototype for a process usefully explored analogically by imagining natural selection as artificial selection writ large. This casual – what we now call 'figurative' – use of metaphor has been a Darwinian legacy amply on display in the writings of Richard Dawkins, who thinks he can write of 'selfish genes' without imputing selves to genes, and a 'blind watchmaker' without implying the existence of a watchmaker, let alone a blind one. Darwin's own casual use of metaphor baffled his contemporaries. Generally speaking, the historical trajectory of a metaphor in science can proceed in one of two directions: either it is replaced by a radically different but literal account of what the metaphor figuratively stood for, or the metaphor itself comes to be elaborated as an analogy whose implications are explored empirically, resulting in a more-or-less literal interpretation of the metaphor.

The former trajectory is familiar from the replacement of spiritual with mechanical explanations in the histories of biology and psychology. What first appear to be autonomous agents turn out to be the functioning parts of some larger system; thus, the competing demons in our soul turn out to be alternative ways in which our brain processes data. (Indeed, a 'pandemonium' model of cognitive processing was popular among psychologists in the 1970s.) The limits of this approach are worth noting. In particular, the parameters of the system must be first identified by an agent outside the formal context of inquiry. Behaviorism worked well in psychology as long as the psychology of the behaviorist was not itself under investigation. Otherwise, assumptions and motives would be revealed that, had they been part of the data processed by the behaviorist's subjects, might well have led to their behaving differently.

The latter trajectory is exemplified by the ancient metaphor of atoms as the ultimate but invisible constituents of matter, which only came to be empirically vindicated in the early 20th century. Because atoms were allegedly material entities that nevertheless managed to elude empirical detection, their existence was widely doubted in the scientific community – that is, until the postulation of their existence became the only reasonable explanation for quantum phenomena. The founder of modern genetics, Gregor Mendel, envisaged the metaphor of natural selection along similar lines. His Christian beliefs led him to suppose that a systematic experimental study of artificial selection would help explain – and improve – natural selection. In the course of his inquiries, Mendel postulated atoms of inheritance, what he called 'elements'; the basis of the modern concept of genes.

Darwin legendarily blanked Mendel for a variety of reasons, not least his own poor German and maths skills. But his incomprehension was probably also related to a deep scepticism that the mechanisms of heredity could be ultimately fathomed. Nevertheless, the so-called Neo-Darwinian synthesis rests on the empirical fecundity of Mendel's faith, which ultimately enabled naturalists and lab-based biologists to see themselves as ploughing parallel furrows in the same field. Not surprisingly, this unity of purpose, which effectively rendered natural selection a literal extension of artificial selection, was first achieved by another devout Christian, Theodosius Dobzhansky.

Of course, Mendel's work inspired the promotion of eugenics in the 20th century, not least in Nazi Germany, where the greatest atrocities were committed in that discipline's name. However, Darwinism played a crucial role in the spin given to Mendel's work; Darwinism enabled the Nazis (among others) to avoid taking personal responsibility for deciding who was fit to live and die by portraying eugenics as simply a matter of following nature's orders, a slight personification of natural selection.[42] Thus, artificial selection became less the *intelligent design* than the *blind execution* of natural selection. This point is significant, given the revolution in molecular biology that has occurred in the 60-plus years that separate us from Hitler. We are now in a position to revisit the question of eugenics with fresh eyes, without having to hide behind the 'blindness' of Darwinian natural selection.

Artificial selection now occurs at multiple levels of biological reality and at various points in the life cycle – not only, as Mendel had thought, in terms of breeding partners but also as a conceived organism is gestating or even after it is born. It is a testimony to Darwin's dominance over Mendel in the Neo-Darwinian synthesis that these various interventions are presumed to be instances of *natural* selection, despite their clear basis in human artifice. In any case, this process is often casually depicted as our capacity to produce 'designer babies', a phrase that overestimates the security of the outcomes. However, what *is* true about this depiction is that our personal responsibility for the future of life on the planet, however imperfect our knowledge base, has never been clearer. For the biblical religions, our status as human beings created *in imago dei* hangs on this point. To conclude *either* that we should not intervene at all in the life cycle *or* that individuals should be left to intervene as they wish is to deny our God-given capacity for free will, the source of our personal responsibility. I say 'God-given' to convey the idea that if humans are indeed the crown of creation, then the opinions of *all* humans are of value in any particular decisions about offspring. To be sure, such an ideal can never be fully realised in the world of practical politics. But the ideal will have served much of its purpose if societies refuse to adopt a default stance of either allowing or forbidding artificial selection in the increasingly exact sense that molecular biology has made possible.

CHAPTER 4

Is Intelligent Design Any Less Science than Evolution?

There are at least three ways of being 'not-science': 'anti-science', 'dead science' and 'pseudoscience'. Does ID – or evolution, for that matter – fall into any of these categories? The answer in all three cases is the same; it depends. Let us take them in increasing order of difficulty, starting with what appears to be the easiest case: antiscience.

Antiscience – or Simply Anti-Establishment?

There are two grounds for concluding that creationism is antiscience. One is purely historical and the other, as it turns out, more scientific. The end of the First World War witnessed a revival in the sort of Christian fundamentalism associated nowadays with 'Young Earth Creationism' (i.e. the idea that the Earth's age can be determined from the genealogy of the biblical patriarchs, which would make it about 6,000 years old). However, this movement did not occur in a cultural vacuum. It was 'antiscientific' in the same sense as parallel secular movements in the 1920s that blamed the unchecked growth of science, and especially its metaphysical principle of materialism, for the unprecedented devastation brought about by such military innovations as aerial bombing and poison gas, which characterised what was originally called 'The Great War'. This general step back from a universal natural science in the sense that Newton would have recognised has been called 'reactionary modernism'.[1]

It is worth observing that the link between scientific know-how and military might was not a figment of the reactionary imagination but had been explicitly forged by the supposedly progressive German scientific community, who predicted that their intellectual superiority would translate into victory on the battlefield. It is therefore significant that the original political leader of the US neo-fundamentalists and lead prosecutor at the Scopes 'Monkey Trial', William Jennings Bryan, earned the respect of pacifists everywhere, not least Christians, when he resigned as Secretary of State in 1915 after President Wilson adopted a belligerent attitude towards the Germans. This is the light in which Bryan argued, a decade later against Scopes, that Darwin corrupted human morals by normalising the occurrence of war as part of the 'struggle for survival' that marks us as merely one of the animals.[2]

Germany's defeat was thus widely interpreted as comparable to the arrogance that led to the 'Fall' recounted in Genesis, whereby the first humans stood guilty of presuming entitlement to knowledge that only God possessed. A secular variant of this antiscientific reaction may be found in Martin Heidegger, a former theology student who, like the neo-fundamentalists, harked back to Martin Luther's suspicion of all claims to expert authority in matters of the spirit, be their source the papal seat or the Pope's scientific challengers such as Copernicus, whose sun-centred universe contradicted not only the Bible but the evidence of ordinary human eyes. The Luther-inspired suggestion was that claims to higher expertise swung the pendulum of original sin to the opposite extreme, pre-empting the free judgement that humans retained as a result of having been created in the image and likeness of God. Thus, the Christian neo-fundamentalists and the Heideggerian existential phenomenologists, united in their reactionary modernism, called for a removal of artifice and a return to 'nature' (aka ecological balance), including a celebration of agrarian virtue. The upshot of these movements was, generally speaking, populist and anti-intellectual.

The second ground for deeming creationism antiscientific is simply that it dissents from the dominant scientific opinion concerning the nature of life, largely because it interprets and weighs the biological evidence differently. Of course, this difference is motivated by the probative value that creationists assign to the Bible as history. But they are also emboldened to interpret and weigh the biological evidence differently by the rather different interpretations and weightings that evolutionists themselves give of the bodies of evidence that are deployed to corroborate the Neo-Darwinian synthesis.

Take the so-called molecular clock hypothesis, whereby the time elapsed from when two species diverged from a common ancestor is measured in terms of DNA differences, themselves taken to be products of mutations that have been naturally selected. It follows that greater DNA differences mean earlier evolutionary divergence. For those who come to evolution by way of molecular biology, like DNA co-discoverer James Watson, the molecular clock hypothesis provides an excuse to ignore fieldwork, the results of which – via the radiometric dating of fossils – provide new information only about the *timing* but not the *ordering* of species in the great phylogenetic tree of life, whose structure can be discerned from DNA alone.[3] This tendency to ignore fieldwork is reinforced by some Orwellian turns of phrase, not least 'fossil gene', which refers *not* to the genetic material left on the fossils of extinct organisms, but to genetic material present in extant organisms that lacks adaptive value and hence is presumed to be the remnant of an ancestor.[4]

Unsurprisingly, palaeontologists and field biologists do not accept this subordinate position. They see themselves as empirically testing the molecular clock hypothesis, not simply assigning time-checks to its presumptively true logical consequences. One especially principled school of taxonomic biology, the cladists, goes so far as to refuse to grant the existence of a common ancestor between any two species until evidence of that ancestor has been actually discovered in the field or the fossil record.[5] So, it seems, neither side of the lab–field divide in biology wants to be left with the task of timing evolution. Richard Dawkins' often repeated claim that evolution would be falsified if fossils of large mammals were found in Cambrian-era rock is a red herring. Even if no such mammals were ever to be found, that would still leave open the question of whether the Cambrian explosion happened 5,000 or 500,000,000 years ago, which is of at least as much concern to creationists as the order in which creation unfolded. As it happens, evolutionists outsource that crucial part of the story to physicists and the vicissitudes of the radiometric dating of trace elements in fossils.

Even though creationists rhetorically benefit from such dissent in the biological ranks, their scepticism – or open-mindedness – is epistemologically legitimate as a response to two perennial conflicts in scientific methodology. One is between deductive (from theory to data) and inductive (from data to theory) approaches to the justification of scientific reasoning. These are represented by the molecular clock hypothesis and the cladist method, respectively: which is ultimately accountable to which? The other conflict is between 'quantitative' (i.e. lab-based) and more 'qualitative' (i.e. field-based) approaches to the conduct of science. It was precisely conflicts of this general sort that prevented the unification of the social science disciplines in the 20th century. I shall return to this point later in the chapter. In this context, molecular biologists are like Marxists who read the past and the future from the present, seeing both atavism and potential in contemporary social formations. In contrast, the cladists are like historians and anthropologists who interpret evidence as referring exclusively to the time and place of its production, without the Marxists' echoes and foreshadows.

Dead Science – but Resurrectable?

Turning now to the intermediate case of 'dead science', we encounter the coinage of Philip Kitcher, the great dampener among contemporary philosophers of science, whose reputation rests largely on his ability to contain excessive responses to Neo-Darwinism, from both sociobiologist zealots and creationist sceptics. As the field of play has shifted over the years, so too has his rhetoric. Thus, contrary to his earlier views, Kitcher nowadays claims that ID is not pseudoscience but 'dead science'.[6] With this phrase, he concedes creationism's utility to biological inquiry in the past, but insists that it has now outlived its usefulness and should be discarded in the name of scientific advancement. In other words, what neo-creationists continue to assert might have been reasonable once, but not any more. Kitcher concludes that, strictly speaking, today's ID belongs with science's 'undead' because its proponents remain convinced that their lines of inquiry are still alive, despite the rightful protests of evolutionists that they died long ago.

That Kitcher means 'dead science' pejoratively is self-evident. Nevertheless, neo-creationists, not least ID theorists, can easily re-spin the phrase to draw attention to the centrality of history to a proper understanding of the nature of science. Expressions like 'dead language' and 'dead species', the latter ironically connected to Darwinism's own normative horizons, provide an alternative semantic frame of reference. They carry connotations of a diminution in life's overall diversity and a reduction in the opportunity for other languages and species to enrich themselves through symbiosis. The underlying suggestion in both cases is that death was avoidable. And as long as an historical trace remains – in the form of codified grammars or recovered DNA strands – the language or species, at least in principle, can be resurrected and perhaps even reincorporated into a supportive ecology. The most striking case of a language enjoying this fate in modern times is Hebrew, which had long been used only for religious purposes until it was revived as an ordinary written and spoken language in the late 19th century. In the case of recovered species, *Jurassic Park*-style nightmares immediately spring to mind, but the reality is that for species extinct for tens of thousands, rather than millions, of years a sufficient amount of intact DNA may eventually enable them to return to life.

So-called 'dead science' is no different. Indeed, the travails undergone by the concept of evolution, not least Darwin's version, illustrates the ever-present prospect of intellectual resurrection. At the 50th anniversary of *Origin of Species* in 1909, Darwinism was regarded as Freudianism is today, more multivalent ideology than testable hypothesis, with 'adaptation' akin to 'unconscious' as a free-floating device for the telling of endless just-so stories of why things work when they do, and why they don't when they don't.[7] As we shall see below, this judgement continued to colour Popper's early view of Darwinism as pseudoscience. Before Darwinism had been embraced by late-19th-century lab-based biologists like August Weismann, who devised experimental means of distinguishing an organism's inherited and acquired qualities, evolution was widely seen as an irrationalist doctrine committed to a metaphysics of life as a 'vital impulse' whose understanding escaped normal scientific strictures. Such an irrationalist sense of evolution could be found in the ancient atomists and the Eastern world religions. It famously worried T.H. Huxley, who eloquently argued that our humanity lay in organised resistance to natural selection, but evolution's irrationalism was precisely what attracted Freud to Darwin. However, as the mechanisms of inheritance came into clearer view, culminating in the discovery of DNA, 'evolution' came to stand for the ultimate explanation of the changes undergone by those mechanisms in real time.

The finality implied in 'dead science' may itself turn out to be what Paul Ricoeur called a 'dead metaphor', in this case a metaphor based on a complacent and increasingly obsolete notion of 'death' itself. Death is slowly but surely passing from fate to choice. Consider the increasing meaningfulness attached to the phrase 'letting die'. Modern medicine has already removed much of the fatalism that traditionally surrounded death in ways that allow us to anticipate and hence participate in the termination or extension of life – not least, in the latter case, by supplementing or replacing body parts.

At a more abstract level, contemporary analytic philosophy, ranging from Wittgenstein's philosophy of language to Quine's philosophy of science, has promoted the idea that one's theoretical horizons are always 'underdetermined' by the default interpretation given to current evidence. In other words, there is always logical space for taking any such agreed body of evidence as a basis for moving inquiry in a new direction.[8] Thus, promoting creationism via ID might be analogous to resurrecting an extinct organism or a defunct language in an environment where it can now flourish by drawing on the same resources as other extant organisms and languages. It is perhaps not by accident that the originator of this 'underdeterminationist' line of thought was the Catholic physicist Pierre Duhem, who wanted science to enjoy unfettered inquiry without pre-empting the resolution to ultimate metaphysical questions, which is arguably the level at which the dispute between evolution and creation matters the most.

All of this requires that creationists recover their ground as a scientific movement with a claim to the same body of evidence as the evolutionists. That the history of biology is deeply indebted to creationists – from Linnaeus' conception of species to Mendel's conception of hereditary factors – automatically gives a neo-creationist theory like ID more *prima facie* claim to scientific legitimacy than, say, the 'flying spaghetti monster', a derisory image of a placeholder for any theory that postulates an intelligence behind the nature of life that beats the odds of natural selection. At the same time, the leading centre in the US for the promotion of ID, Seattle's Discovery Institute, has recently published *Explore Evolution*, a textbook that exemplifies the desired underdeterminationist viewpoint to a tee.[9]

Pseudoscience – or Science Trying to Do Too Much?

Until the recent emergence of 'antiscience' and especially 'dead science', as used specifically against ID, 'pseudoscience' was the philosophical put-down of choice for creationists. In other words, creationists are either making false scientific claims or dressing religion up as science. But when the term was first used in the 1930s by the logical positivists and their fellow traveller Karl Popper, it referred to the premature closure of inquiry by an inappropriate extension of scientific authority. In this context, science lost its empirical character and became reprehensibly 'metaphysical'. Thus, Popper applied the label 'pseudoscience' to a broad class of theories he labelled 'historicist', according to which aspects of the past overdetermine the future – be it of one's own life, one's society or even the entire human species.[10] Such theories include astrology, psychoanalysis, Marxism and evolutionism, including Darwinism. A diverse bunch, to be sure, yet they share not only a strong grounding in what has happened but an even stronger conviction that certain things will happen. The result is a counsel of fatalism that, depending on circumstance, may take a quiescent or a violent form, ranging from suicide to revolution.

The inclusion of Darwinism in this list of epistemic offenders has been the source of much controversy from the outset, with matters only complicated once Popper removed Darwinism from the list in the 1960s. In Popper's defence, it is worth recalling that the 'Darwinism' he dubbed pseudoscientific in the 1930s was not quite the same theory he rehabilitated 30 years later. In the interim, general agreement was reached on the 'Neo-Darwinian synthesis', which brought the lab-based experimental sciences of genetics and molecular biology to bear on Darwin's own natural history of the evolution of species. The synthesis upgraded Darwin's theory of natural selection from a just-so story that provided *post hoc* explanations for any and all aspects of a surviving organism, to a blueprint for testable accounts of how specific genetically variable organisms differentially reproduce in changing environments.

A subtle feature of the Neo-Darwinian synthesis is that the aspects of Darwinism that were most vulnerable to the charge of pseudoscience – its anthropological implications – were removed from the core of the synthesis. Darwin himself had used the same tactic in *Origin of Species*, which explicitly refrained from arguing that *Homo sapiens* is also the product of natural selection. His more forthcoming views in later works, notably *The Descent of Man*, were

– and still are – deemed politically incorrect, despite the greater relevance of these works to contemporary sociobiology and evolutionary psychology. Unsurprisingly, the phrase 'Social Darwinism' was coined only in 1944 to suggest that any straightforward application of Darwin's theory to human affairs was an ideologically motivated pseudoscientific distortion – *and always had been.*[11] The timing of the coinage tracked a clear distinction between 'scientific' Darwinism and 'pseudoscientific' social Darwinism that had only recently emerged in the mind of Hofstadter and his contemporaries. It was *not* present in the figures he studied, who had flourished half a century earlier, in the immediate afterglow of Darwin's original work. Darwin remained silent on human matters in 1859 in order to avoid theological controversy over the loss of our privileged place in nature. But nearly a century later, in the wake of the Second World War, Neo-Darwinian reticence about humanity made sense as a backlash against Nazi eugenic policies that had claimed a basis in Darwin. However, not all Neo-Darwinians cooperated. The biologist who actually coined the phrase 'evolutionary synthesis', Julian Huxley, has been written out of most histories of his discipline because he still believed that eugenics, albeit administered with a socially progressive hand, was a natural extension of the synthesis.[12]

In his typically cagey fashion, Popper did not exactly recant his earlier appraisal of Darwinism. Rather, he upgraded his opinion of its 'metaphysical' character. In philosophical jargon, Popper shifted the status of Darwinism from the 'context of justification' to the 'context of discovery'.[13] In other words, the true value of Darwinism is to be found less in the actual truth of its claims, which can be interpreted in any number of ways, than in the inspiration it provides for more specific empirically testable claims. In this respect, Darwinism is a powerful aid to the biological imagination, on a par with, say, the atomic theory of matter. However, this is an argument only for the inclusion of Darwinism in the science curriculum, not for the exclusion of alternative metaphysical frameworks that may prove equally inspirational. Indeed, Darwinism's superiority would be demonstrated only if it inspired novel correct predictions that contradict those of competing theoretical frameworks.

This is an exceptionally tall order for any scientific theory to meet as grounds for asserting epistemic superiority. Neo-Darwinism and ID predict much the same thing to happen in both nature and the lab at the level of empirical observation. This explains how ID can be couched entirely in the language of science and draw its leading defenders from the ranks of the scientifically trained – albeit in the more mechanistic and applied sides of science (e.g. biochemistry and bioengineering). The intellectual differences between the two theories emerge in terms of the explanations they offer for what they jointly observe. These explanations are difficult to compare with each other because they interpret the same data in radically different theoretical frameworks. This 'incommensurability', to recall Thomas Kuhn's once-fashionable term, can be captured in how each defines the burden of proof. Consider adaptations, defined as physical structures that facilitate the survival of a form of life.

Neo-Darwinists presume that adaptations (however specified) come about by means unrelated to whatever functions the relevant physical structures currently serve.

This presumption minimises the role of forethought in the organism's constitution. In contrast, ID proponents presume that adaptations (again, however specified) reflect forethought that, together with knowledge of the medium in which such thought takes shape, explains the relevant physical structures. On both sides of this interpretative divide, there is plenty of room for divergence on exactly what counts as an adaptation. For example, Neo-Darwinists look weak when organisms appear purpose-built, while ID proponents look weak when organisms appear jerry-rigged. Of course, both sides can finesse their complementary weaknesses. Neo-Darwinists are happy to stretch evolution's time frame for as long it might take for chance-based processes to consolidate into stable adaptations, while ID proponents are free to shift point of view, so that physical structures that appear sub-optimal when regarded in their own right, turn out to be optimal once seen from the wider or narrower 'divine' perspective.

In terms of the recent controversy surrounding ID, these dialectical manoeuvres acquire a special poignancy. ID's most fearless champion, the US biochemist Michael Behe, rhetorically backfired when he claimed that Neo-Darwinism was incapable of explaining the 'irreducible complexity' of the cell, whose purpose-built character resembles a mousetrap, none of whose parts can function independently. In response, Darwinists cleverly played both sides of this dialectic by deftly shifting between natural history and laboratory artifice. On the one hand, they postulated a sufficiently long time frame during which the cell could have emerged in an entirely self-organising manner without forethought; on the other, they showed that parts of this supposedly purpose-built entity can function perfectly well in other capacities in contexts conjured up in the laboratory. Of course, the Neo-Darwinists themselves cannot demonstrate the exact evolutionary path(s) by which the cell *historically* came to have the purpose-built properties it appears to have. But all they had to do to defeat Behe was to show that such an explanation is *possible* within the Neo-Darwinian framework by conjuring up multiple hypotheses with the aid of laboratory experiments and computer simulations, all of which are ultimately speculative historical re-enactments.

Perhaps Behe should not have taken Darwin's bait: 'If it could be demonstrated that any complex organ existed which could not possibly have been formed by numerous, successive, slight modifications, my theory would absolutely break down.'[14] An elementary course in the rhetoric of science would have taught Behe that all arguments from impossibility in science are doomed to failure: they always end up revealing the arguer's lack of imagination. Darwin's so-called challenge is best read as a rhetorical flourish, since you can't prove that something is impossible unless its existence would amount to a logical contradiction. In that sense, Darwin has already won his own bet. And even if 'impossible' is taken to mean 'empirically very improbable', Darwin still has the cards stacked in his favour, since he presumed a geological age of the Earth of at least several million years – the longer the better, from the standpoint of allowing chance-based processes the time to have evolved into stable structures.

Behe would have been better advised to concentrate on a different modal argument: *that something is possible does not mean that it is actual, let alone necessary.* As long as evolutionists cannot bridge the modal gap between the possible and the actual in their core domain, the natural history of the Earth, the conceptual space remains for alternative explanatory scenarios for the emergence of the cell and other *prima facie* intelligently designed features of nature.

The persistence of this gap suggests a deeper confusion in Neo-Darwinist thinking that may reflect the need to keep multiple constituencies singing from the same hymn sheet. In the jargon of the philosophy of social sciences, Neo-Darwinism confuses *nomothetic* and *idiographic* inquiry, the study of recurrent tendencies and the study of unique events. The former is conducted in the lab, the latter in the field, yet evolutionists routinely elide methodologically significant differences between these rather different data-gathering sites – sometimes from one sentence to the next. Here are two examples:

> Modern biologists usually point to examples of observable evolutionary change in the short term, together with the kind of similarities seen between species that Darwin discussed and that suggest common ancestry. For instance, we can see the evolution of drug resistance in HIV (the virus that causes AIDS) within two to three days in an AIDS patient.[15]

> The human immunodeficiency virus contains in its brief history the entire argument of *The Origin of Species*: variation, a struggle for existence, and natural selection that in time leads to new forms of life. Geography tells part of its story, as do fossils, and its genes are a link to distant relatives with which it shared an ancestor long ago. They reveal

> a hierarchy of order as evidence of descent from a common source pushed further and further into the past.[16]

The invocation of God as cause seems weakest (or most superfluous) when evolution is presented nomothetically; say, via repeatable experiments or simulations that push offstage any sense of the miraculous from the generation of life. (The 'only' miracle is that the experiments and simulations themselves occur.) Yet, stripped of its metaphorical suggestiveness and analogical applications, Darwin's theory of natural selection is supposed to explain the *actual* history of life on Earth. It is the unique nature of that account that keeps the inference from the nomothetic to the idiographic open, and with it the possibility of divine intervention at specific junctures. From a social scientific standpoint, the appeal to divine intervention in natural history is no more preposterous than what Thomas Carlyle promoted in the 19th century as the 'great man theory of history', which supposed that events would not have turned out as they did had particular individuals not acted as they did at crucial points. To be sure, Carlyle's sense of history hardly commands a consensus, but it is still deemed worthy of dispute.

In effect, Behe was stymied by an equivocation in ordinary language on what it means to say that one 'can explain' something: sometimes the charge can be met simply by the sort of *possible* explanations that evolutionists have provided in abundance. Thus, whenever Behe claims that a cell, organ or organism could not have evolved bit by bit over a very long time, because its intermediate 'incomplete' versions would have lacked the adaptive capacity to survive another generation, his nemesis Kenneth Miller converts the claim's topic from natural history to experimental demonstration. In the process, Miller does not actually show what Behe says cannot be shown. Instead, he shows how today's scientists can simulate in the lab what modern evolutionary theory presumes to have happened in the past without the intervention of the scientists themselves – or, more to the point, God. In short, Miller provides an actual model of a possible history. The rhetorical import of Miller's response is to leave the impression that even if Behe is eventually proved correct in his claim that natural selection does not explain how the cell actually came to be as it is, in the short term he appears to be pre-emptively excluding a demonstrably possible account of the cell's emergence.

Nevertheless, a cynic with a social science background might reasonably conclude that Miller has pulled a methodological bait-and-switch. The social sciences enjoy an epistemological privilege in this discussion because they have most rigorously addressed the complex of issues implied here: how are we to relate together the findings reached by multiple methods that are meant to be applied to settings rather different from the ones in which the knowledge was first obtained? We might be interested in knowing about the remote past (e.g. the 'origin of species' in biology) or what is likely to work in the future (e.g. the prospects for eradicating a disease or preserving a species), but any knowledge we acquire, by whatever means, is in the so-called extended present. Some of it is acquired in environments that involve relatively little physical intervention from the scientist (e.g. observing nature 'in the wild'), others involve considerable prior intervention (e.g. 'observing' nature in the lab), if not *ex nihilo* creation (e.g. virtual observations made on computer simulations). In each context, the scientist may influence research outcomes at various points and in various ways, both in terms of what happens and what is seen to have happened. There may also be differences that remain unexamined in the observed and the target populations that are instrumental in determining, respectively, the actual and hypothesised events.[17]

To feel the full force of these methodological challenges, suppose an experimental psychologist impressed with the explanatory value of Marx's historically based theory of class conflict wanted to see if it could be simulated in the controlled setting of the laboratory. He or she would thus manipulate some of the variables that Marxist historians have found salient in determining the shape and direction of class conflict – for example, by informing subjects of each other's class positions. Now suppose further that the outcomes of these experiments conform to Marxist expectations. What follows? Has the psychologist captured some recurring historical dynamic or even some timeless element of human nature? Reciprocally, do these experiments increase the credibility of the Marxist historical accounts? To be sure, *some* social scientists would take the psychologist's experimental outcomes as reinforcing the Marxist account. The mix of methods would not trouble them, at least on balance. These social scientists would probably include economists and sociologists who see their work as closer to the natural sciences than the humanities. They would agree with most Darwinists on what methodologists call 'uniformitarianism', the idea that our default scientific attitude should be that the processes that normally occur now are the same ones that have occurred throughout history – that is, indefinitely in both the past and the future.

However, what exactly follows from the uniformitarian imperative differs depending on whether it is enacted in the field or the lab. In particular, the regularity of naturally occurring events need not coincide with that of those produced artificially on a regular basis. This is a more concrete way of capturing the idea that correlation is not causation. Generally speaking, the lab scientist controls which potential causes are allowed to be operative, whereas the field scientist usually cannot. Regularities that look similar but are drawn from markedly different settings do not necessarily reflect the same underlying causal structure. In this respect, regardless of their substantive views on the evolution–ID debate, social scientists can perform a valuable service simply in questioning the methodological assumptions made by evolutionists as they glide effortlessly between data gathered from radically different sources.

Social scientists are familiar with this facile elision of methods from the likes of E.O. Wilson and Steven Pinker, who routinely pick and mix from social science research without regard for the theories and methods of the original researchers, as long as they can be made to support a sociobiology or evolutionary psychology agenda. Such a lack of methodological scruples applies equally, if not more so, in evolution's natural scientific heartland, where the findings are not based on human subjects. However, the evolutionists themselves are destined to confront this unscrupulousness, as they increasingly discriminate between animals of the same species in terms of psychological and sociological variables that are traditionally reserved to register human differences. We might regard this development as ID's revenge, since the idiographic method itself was originally justified by each human's possession of a unique consciousness, or 'soul'.

A good example of evolutionary theory's methodological unscrupulousness is provided in one of Miller's putative refutations of Behe. It involves the attachment of the human growth hormone to cell receptors, which resembles the relationship between a key and a lock. For Behe, to tamper with the shape of either should lead to malfunction. For example, if the receptor loses one of its amino acids, then the hormone should not be able to attach to it. Miller characterises the relevant experiment as 'a group of scientists in California decided to watch the evolution of a new interface between two proteins'.[18] Yet this formulation already misleads on the issues that divide Miller and Behe. Miller presumes throughout his discussion that 'evolution' refers to the same process(es) regardless of whether the evidence comes from the field, the fossil record or the laboratory.

Moreover, Miller exaggerates the passivity of the scientists who 'watch' evolution, when his own detailed account of the scientists' artifices makes clear that they are more than merely watching evolution: they are creating it. To 'let evolution take over', in Millerspeak, is for scientists to engage in elaborate experimental stage-setting. It is *they* who randomly mutate the coding regions for five amino acids in the growth hormone, generating roughly 10 million different mutant combinations, which are in turn tested ('selective pressure is applied', in Millerspeak) to see which, if any, of them bind to the previously mutated receptor. It turns out that a version of the hormone is generated that fits the mutated receptor nearly 100 times tighter than the nonmutant version. Instead of crediting this success to a triumph of human ingenuity, Miller concludes the following:

> [It] illustrates how two proteins, two parts of a biochemical machine, can evolve together. A study like this shows that evolution can act in unexpected, unanticipated ways to fashion novel proteins. And it shows … how protein-to-protein interactions are maintained. So long as selective pressure, even slight selective pressure, exists, it will hone and refine the interactions between two proteins of any biochemical machine.[19]

Yet, it was the scientists who first induced the mutations and then manipulated the selection environment to see if the desired fit between hormone and receptor could be re-established. What is 'unexpected' and 'unanticipated' here is twofold: that a new fit was found, and the exact fit that was found. Whether this represents a process that spontaneously occurs *outside* the laboratory is not directly addressed by the experiment. It is simply presumed to happen throughout nature. What is clear, however, is that considerable frontloading – what could reasonably be called 'intelligent design' – on the part of the researchers was necessary in order to reach the conclusions they did. For example, one would like to find out more about how the 'random mutations' were generated in the laboratory. A properly-conducted 'trial-and-error' experiment does not itself arise from trial and error, but only after some elaborate stage-setting in the laboratory that strategically limits the range of possible causes and effects. At that point, then, yes, the outcomes are 'unexpected' and 'unanticipated'. Under the circumstances, it is unsurprising that Behe would characterise science as seeking 'the edge of evolution', that is, the point or level at which natural phenomena can be truly explained in terms of some process that had not been premeditated *at all*.[20]

So, does the experiment Miller cites really refute ID? To an ID theorist, all successful laboratory demonstrations of evolution attempt to simulate on a small scale God's own world-creating methods. These involve controlling certain conditions and allowing others to vary, both by acts of will. The extent to which the human simulations approximate to divine creation may be measured by their generalisability to situations outside the laboratory in so-called 'real life' or '*in vivo*' settings. It is here that the gap in knowledge and power between the human and the divine is most keenly felt. But if the gap can be narrowed over time – in such a way that the artifices of the laboratory can be increasingly used to turn nature to human ends – then the ID theorist is justified in concluding that scientists are coming closer to grasping the divine creator's methods. Certainly, the histories of both natural and social science offer grounds for at least modest optimism. Admittedly such an argument, when made so baldly in the past, has infuriated theologians at least as much as scientists. In recent years, much of the fury has centred on the extension of medicine beyond therapeutic aims to the outright enhancement of the human condition. In a phrase: *playing God*.[21]

Had Miller advanced his critique of Behe a century ago, he would have been called a 'vitalist' and Behe a 'mechanist'. The two philosophies underwrote competing schools of developmental biology, both of which had theological overtones and neither of which were especially friendly to Darwin.[22] Mechanists and vitalists located God rather differently. In the Kantian philosophical jargon, the mechanist's deity *transcends* nature, whereas the vitalist's deity is *immanent* in nature. For mechanists, divine intelligence exists outside – and hence is imposed on – nature, notably through a sequenced pattern of forms, or developmental stages, each of which lays the foundation for the next. This way of thinking has perhaps persisted longest in child development studies under the spell of Jean Piaget. In contrast, vitalists treat nature as possessed of its own intelligence – or *élan vital* ('vital force') in Henri Bergson's famous phrase – that pursues its ends by whatever happens to be the available means, even in the face of adversity. Thus, an organism whose normal development is disturbed will compensate as much as possible to reach for a fully mature state. Miller reveals his vitalist scruples in what he calls the 'true acid test' of Behe's hypothesis of irreducible complexity, namely 'by using the tools of molecular genetics to wipe out an existing multipart system and then see if evolution can come to the rescue with a system to replace it'.[23]

Needless to say, 'evolution' always manages to come to the rescue, since for Miller the word means little more than whatever natural processes are presumed to have transpired, should the missing part re-emerge in the course of the experiment. He might as well have invoked *élan vital* as a synonym for 'evolution'. Here Miller is helped by a studied ambiguity in evolutionary concepts that is exemplified in the bacterial flagellum, the explanation of which has been a focus of disagreement between Behe and Miller, forming the centrepiece of Miller's star turn in the witness box in *Kitzmiller*. The flagellum is the whip-like structure that enables bacteria to sail through liquids in the manner of an outboard motor. But does 'flagellum' define a specific physical structure or simply a function that may be performed by a variety of physical structures, each of which may also perform other functions as parts of different organic systems? The mechanist presumes the former, the vitalist the latter. Modern evolutionary theory is ambiguous with regard to the two positions, though clearly vitalism provides the evolutionist with greater interpretive flexibility.

This ambiguity also lies at the heart of the quarter-century public feud between Richard Dawkins and Stephen Jay Gould, which came to an end only with the latter's death in 2002.[24] The difference between Dawkins and Gould is epitomised in a pair of contrasting slogans: 'Form follows function' and 'Function follows form'. The former characterises evolution as 'adaptation', the latter as 'exaptation', a Gould coinage. The difference may be put even more starkly: Dawkins, like Miller, tends to conflate form and function, whereas Gould radically separated them. Evolutionists who, like Gould, tend towards a literal reading of Darwin, distinguish between the *homology of forms* and the *analogy of functions*. These constitute two ways of organising biological knowledge.[25] Homology focuses on the similarity of structure in organisms, analogy on the similarity of function. The former focuses on common physical structures shared by organisms as evidence of a common evolutionary ancestor that also possessed the same structure. But the evolutionary function performed by the same bone or DNA strand in these organisms may have been different, depending on the immediate environment and assuming the structure served any function at all. Indeed, Darwinians feel especially fortified against design-based thinking if the function – or, better still, functionality – of a structure can be shown to have shifted over evolutionary time.

Gould was quite clear about all this, which led him to doubt that the history of life on Earth is well-enough designed for the idea of a creative deity to be justified in the first place. By showing how currently adaptive features of organisms emerged as neutral by-products of other features, which themselves arose as genetically transmitted by-products, he proved himself the true heir to David Hume's sceptical arguments against design in nature. Natural history appeared to Gould as a series of lucky accidents that would probably not happen in the same way again, given a second chance. In contrast, Dawkins, despite his self-avowed 'intellectually fulfilled atheism', has quite happily helped himself to design-based language, not least 'selfish gene' and 'blind watchmaker', to cite the titles of two of his books. In his hands, 'adaptation' is a secular synonym for 'design', and 'natural selection' a secular synonym for 'God'. For Dawkins, the debate between evolutionism and creationism boils down to whether or not God is needed to explain something that can be explained by purely naturalistic means. But unlike Gould, and probably Darwin himself, Dawkins believes that there is something worthy of explanation. That Dawkins has been much more influenced by recent developments in genetics and molecular biology than Gould, whose competence and interests coincided more closely with Darwin's own field-based inquiries, probably underlies his more upbeat design-based version of 'Neo-Darwinism'.

Evolution – the New Astrology?

If the history of science can be said to possess a collective unconscious, then the memory it has tried hardest to repress is that of astrology, which for most philosophers remains the paradigm case of pseudoscience. Yet this is entirely unjustified. Astrology provided the earliest vision of reality as a bounded rational entity, an Earth-centred universe, whose internal workings were to be understood in terms of mathematical equations and its outer workings in terms of deep structures and remote causes. Astrology depicted the physical arrangement of the heavenly bodies as at least constraining, if not outright determining, what takes place on Earth, especially to our own minds and bodies. Astrology's legacy lingers in terms like 'influences' and even 'influenza', which continue to carry the psychosomatic ambiguity of external forces that somehow manage to penetrate and alter our innermost being. But perhaps astrology's most lasting scientific legacy has been the supernatural entity known as gravity, which Newton famously said 'must be caused by an agent constantly acting according to certain laws, but whether the agent must be material or immaterial, is a question I have left to the consideration of my readers'.[26]

Nevertheless, one of the most widely noted and derided moments in *Kitzmiller* occurred when the plaintiffs' lawyer, Eric Rothschild, managed to get Michael Behe to admit that astrology was a scientific theory according to his own definition, which was simply a proposed explanation for a set of facts.[27] The judge also clearly found Behe's admission damning, citing it as a reason for ruling in favour of the plaintiffs. But even more damning was the disrespect for the history, philosophy and sociology of science that the US legal system showed in the process. In the deposition filed before the trial, Behe states that he arrived at his definition after surveying articles citing the word 'theory' in the abstracts published in PubMed, a computer-searchable database of peer-reviewed articles in the biomedical sciences. He found considerable disparity in usage, but concluded that the use of 'theory' to mean a general explanation for disparate phenomena was most prevalent. Behe's point here was that 'theory' is in fact a loose term in scientific discourse, and presumably it is hypocritical for high school textbook authors to pretend to a standard higher than that to which professional scientists hold themselves.

In his cross-examination, Rothschild insinuated that Behe's broad definition of 'theory' was indifferent to matters of truth, and that this led him to infer that even a practice as palpably false as astrology counted as a scientific theory. Had Behe been trained in the history, philosophy or sociology of science, he could have easily met Rothschild's challenge by concluding that all this reveals is the intellectual double standard under which scientific institutions currently operate. The fact that 'theory' can be invoked in the relatively vague sense of explanation when one speaks from inside the scientific orthodoxy, but not when one speaks from outside it – even though in both cases 'theory' has much the same meaning. Similarly, evolutionists are allowed to invoke design language, even when explaining quite detailed research findings, but ID theorists are not allowed the same privilege. A card-carrying evolutionist like Stephen Jay Gould is permitted to question the pervasiveness of adaptation and natural selection in accounting for the origin and survival of species, but creationists are prohibited from turning the same sceptical arguments to their own advantage. Having presented this litany of science's normative inconsistency, Behe could have then pressed the conclusion that to retain its credibility, the scientific establishment either needs to tighten up its own practices or extend the same generosity to more heterodox practitioners.

Interestingly, Rothschild did not go on to offer a theory that conformed to Behe's definition but that failed to make a discernible difference to the history of science. Such theories would include the various creation narratives proposed in cultures outside the orbit of Western monotheism, on behalf of which multiculturalists nowadays argue for a place in the school – though not necessarily the science – curriculum. Had Rothschild raised these narratives, Behe would probably *not* have welcomed them in the science classroom. He might have even reverted to a politically incorrect term like 'myth' to describe them because they – unlike ID – failed to contribute substantially to what we now recognise as science. That admission would have made for a far more interesting, though no less controversial, interrogation. Here, too, I would have been on Behe's side.

In any case, the scandalous inference drawn from Behe's testimony was that at least some theories should be taught as scientific, even if the scientific consensus has deemed them false. Why would anyone in their right mind hold such a view? The answer lies in a lesson from history: the wider the range of phenomena a scientific theory purports to explain, the more likely it will prove wrong in the long term. This is simply because such a theory – be it astrology or modern evolutionary theory – depends on the knowledge claims made by many different sources, each of which may be challenged on grounds quite independent of each other and of the grand theory they purportedly support. The image of a house of cards comes to mind. But notice that the cards can still be played even after the house collapses, which is exactly what happened to the disciplines and practices that contributed to the scientific legitimacy of astrology for 1,500 years. Thus, were serious doubt to be cast on the reliability of the radiometric methods currently used to determine the age of fossils, the time frame of evolution would be placed in jeopardy without necessarily affecting the ordering or classification of species.

Nevertheless, any mention of astrology and evolutionary theory in the same sentence is bound to be met with scepticism. We imagine that astrology in the past was simply a more bombastic version of the New Age charlatanry that passes for astrology today. By the same token, we also imagine that evolution today is a considerably more nuanced and scientifically supported theory than it was in its earlier incarnations as Social Darwinism and eugenics. Both accounts of the relationship between past and present practices in the two fields are exaggerated: astrology has fallen from a much greater height of scientific respectability over the past 500 years, and evolution has made much a smaller conceptual climb over the past 100 years. Astrology was the matrix in which algebra and astronomy were spawned, enabling occult forces like gravity and magnetism to receive their first mathematical treatment. In contrast, most of the key concepts of evolution were in place by the end of the 19th century; hence the ease with which contemporary developments in evolutionary theory can be portrayed as footnotes and amendments to Darwin's original works. The major exception to this claim – the molecular basis of genetics – was developed without any special reference to evolution, although evolutionists have of course turned it to their own purposes.

Beyond its potentially pseudoscientific 'metaphysical' character, evolution also shares a substantive intellectual assumption with astrology; namely, a commitment to the idea that what used to be called 'action at a distance' is responsible for the movement of physical bodies, including our own. Indeed, this idea lies at the core of what is now derided as 'supernaturalism'. Astrology and evolution differ over the *dimension* of this distance: astrology stresses the *spatial* distance of the heavens from the Earth, whereas evolution stresses the *temporal* distance from either our origins or our destiny, the latter typically understood as the realisation of the former's potential.

Where origins are stressed, the result is 'hysteresis', which is what occurs when a long-past event continues to exert a pull on the present, as in the havoc allegedly wreaked on human behaviour by our modern skulls housing a Stone Age brain. If evolutionary psychologists are to be believed, it would seem that nothing has quite managed to reverse the effects of a lucky genetic accident that happened millions of years ago – either 2 or 200 million, depending on whether the relevant part of our brains was born under the sign of ape or reptile in the evolutionary zodiac.[28] Where destiny is stressed, the result is 'teleology', according to which the movement – specifically, the development – of an object is governed by the prospect of the full realisation of its potential. This is typically conceived as a future resting state, prior to which the object will pass through a sequence of intermediate states, having overcome various forms of resistance. Perhaps the most important appeal to teleology in modern biological thought appears in Lamarck, who proposed that all life forms are converging upon a perfected human condition, as the secular realisation of our biblical entitlement to dominion over the Earth.

One now largely forgotten figure who appreciated this connection was James McCosh, the Scottish Presbyterian president of Princeton University in the period immediately following the American Civil War. McCosh spearheaded the first liberal evangelical response to Darwin. He proposed that evolution should be regarded as a 'law' in the same sense as gravity: the one applying to nature over a length of time, the other to nature over an expanse of space.[29] In both cases, one speaks of 'influences' transmitted across the relevant distances – mediated by a substance such as an aether or a germ plasm – that may be somewhat transformed but never entirely eliminated. Put in today's terms, the pull of gravity is comparable to genetic load. However, McCosh himself interpreted evolution more in the future-oriented 'historicist' terms that would later lead Popper to count Darwinism as a species of pseudoscience. McCosh's particular brand of evolutionism was part of a half-logical, half-theological campaign to inhibit people from judging the validity of a proposition on the basis of its immediate consequences. It was not that McCosh denied the relevance of consequences to validity. On the contrary, he believed that validity was tied to ultimate import, something revealed only in the fullness of time. Of main concern to McCosh was the temptation, born of human arrogance, to read long-term significance from short-term effects – to assume that we know more than we really do.[30]

While McCosh's distinctive teleological reading of Darwin encouraged a critical attitude towards the dominant forms of biological adaptation, much of evolution's wider cultural authority, like that of astrology, has traded on a sense of cosmic determinism, if not fatalism, that infers what is meant to be from what is. Yet, as Popper rightly observed, both fields have found it notoriously difficult to make accurate predictions, as they both require precise knowledge of many interacting variables. What is true of horoscopes is no less true of genetic counselling. In the case of astrology, the problems include the accuracy of star charts (the geometric representation of celestial motions), not to mention that of the information provided by the client seeking astrological guidance. In the case of evolution, to determine the extent to which a given trait in a given person is genetically expressed requires a sequenced map of their genome, as well as an accurate account of the environmental conditions to which they have been and are likely to be exposed.

The most popular image associated with astrology, the harmony of the celestial spheres, inspired the quest for a unified account of matter and motion. This quest, a core concern of medieval Islamic science, left an indelible legacy in *algebra*, which literally means 'restoration of balance' in Arabic. Thus, the solution to the multiple functions in a complex simultaneous equation would provide the mathematical expression of this celestial harmony. In this process, now the routine stuff of computer simulations, the algebraist would enact a potential thought of God. Not surprisingly, the leading European contributors to the development of algebra before the 17th century were astrologers, the most famous being Giordano Cardano.[31] However, Isaac Newton, who sympathised with astrology's ambitions, determined that the world order is not quite so self-regulating. But rather than denying that there is a celestial harmony to be fathomed, he found respectable mathematical means by which to license God's periodic interventions to restore the harmony of the spheres. This strategy, though eventually abandoned in physics, inspired the Keynesian idea that governments should intervene to flatten out the phases of the business cycle in order to restore general equilibrium in the economy. As it turns out, John Maynard Keynes was among the first to read Newton's private papers, which detailed his lifelong interests in theology, alchemy and astrology that he deliberately hid from public view in his physics.[32]

The advanced form of astrology practised in Islam made its way into Christendom in the 13th century through the visionary early Oxford scholars Robert Grosseteste and Roger Bacon. Eerily presaging Cold War rhetoric, they were convinced that Christendom's triumph over Islam depended on winning the 'space race' that astrology had opened up, which allowed humans the opportunity to harness cosmic forces in order to chart their own destiny.

In this context, the observatory functioned as the particle accelerator does now. It was an optics-centred world view that reached its fullest expression in Newton's *Principia Mathematica*, which for at least two, and arguably three, centuries has symbolised the West's universalist intellectual and political aspirations.

However, in the four centuries that separated the early Oxford scholastics from Newton, astrology grew in secular importance, resulting in the field's knowledge claims becoming 'unfalsifiable', the specific quality Popper attributed to pseudoscientific theories. In other words, astrologers refused to submit to a public test that might reveal a fundamental error in their theories. Their reasons are familiar to us today. When astrology was within the reach of only the powerful and was used to decide the timing of military campaigns, it was shrouded in secrecy for what we now call 'national security reasons'. But starting in the 15th century, when astrologers moved into Europe's emerging private sector, and personal horoscopes increasingly became their stock in trade, client confidentiality was cited as grounds for refusing to release their track records.

Nevertheless, as astrologers acquired status, they made greater claims to knowledge for their field. Many leading Renaissance intellectuals, including Ficino, Paracelsus and Pomponazzi, tried to leverage astrology's historic significance and burgeoning clientele into a foundational role in the university medical curriculum. Like many enthusiasts for evolutionary biology in the medical profession today, they believed that astrology would finally render their ancient art a genuine science with a deep causal sense of the extent to which humanity's well-being was embedded in the cosmos. This turned out to be a step too far, placing astrology under much sharper critical scrutiny than ever before. Suddenly everyone was a Popperian *avant la lettre*. Challenges to astrology's pretensions, even among fellow practitioners, became very public affairs which only served to cast doubt on the entire enterprise – even when astrologers were shown to have drawn valid conclusions. Thus, while Galileo engaged in largely fruitful correspondence with Johannes Kepler, he doubted Kepler's (correct) lunar explanation of the tides, because its postulation of physically remote causes for earthly events was redolent of Kepler's service as the Holy Roman Emperor's court astrologer.

While all this increased scrutiny eventually undermined the scientific credibility of astrology, it inspired some major breakthroughs in thinking about the historical and scientific method. Here are three examples. Machiavelli's notorious sensitivity to contingency in human affairs, not least the proven capacity of expert politicians to turn the immediate situation to their advantage, was predicated on a deep-seated scepticism about the star-fated explanations favoured by astrologers. Descartes' project of setting knowledge on indubitable foundations emerged from thinking about what must remain true of reality even as astrologers contradicted each other about the relative causal powers of the heavenly bodies on earthly affairs. Finally, King James I's court physician, William Gilbert, distinguished magnetism and electricity as physical forces by trying to make sense of the idea that stellar influence is subject to varying degrees of 'interference' from other, more local, factors. Gilbert's experiments, though not wholly valid in their conclusions, nevertheless provided a concrete backdrop for the original formulation of the scientific method by his fellow courtier, Francis Bacon.

Modern evolutionary theory, as we have seen in these pages, is subject to vagaries of interpretation just as fundamental as those that ultimately floored astrology. Here is a list:

1. Is the overall process of evolution directed or undirected – Lamarckian or Darwinian? If we deliver a mixed verdict, then when and where does the directed yield to the undirected?
2. Is design something that genuinely needs to be explained or an illusion to be dispelled? Again, given a mixed verdict, when and where should forms of life (or parts thereof) be regarded as 'adaptations' or 'exaptations'?
3. What about the so-called 'tree of life'? Depending on whether the tree is arranged on the basis of an organism's surface morphology or its underlying genetic structure, a somewhat different story about the actual course of evolution as 'common descent through modification' is told.[33]
4. The above questions cannot be answered unless the matter of evidence has been resolved: what is the relative weighting of evidence drawn from sources as disparate as radiometric analysis, computer simulation, field observations and laboratory experiments?
5. To what extent is evolutionary history responsible for what we normally regard as human history, and vice versa? How much can environmentally-induced changes in one generation alter the selection environment for subsequent generations?

6. Finally, if natural selection is, in some sense, a 'chance-based process', exactly to which theory of chance does the evolutionist subscribe? Presumably it is more than Aristotle's idea of two or more independent causal processes whose outcomes coincide. Natural selection's grounding in genetics certainly implies a commitment to some idea of statistical frequency. But does it also involve metaphysically deeper notions like 'propensities' and 'dispositions' that might be unleashed systematically under the right experimental conditions? That Neo-Darwinism is taken to underwrite today's bioengineering projects suggests just such metaphysical depth, though it is alien from Darwin's original line of thought.

The fact that the above disagreements remain mostly – though not entirely – confined to academic publications simply reflects the much greater professional control that evolutionists vis-à-vis astrologers have exerted over how their expertise is evaluated.

The history of science should not discourage the search for grand unified theories worked out in exquisite logical and empirical detail, which is after all what distinguishes science from other ways of knowing. Rather, history's role is to sensitise us to the necessary fragility of science's products, and to inspire ambition in those who would propose something grander. Astrology, in its late-15th-century heyday, was a vast and impressive construction that within little more than a century would be discredited for having overreached its grasp. Astrology turned out to be false in the whole but not in its parts, most of which were either directly absorbed by other sciences or eventually mutated into concerns now taken up by other fields that are normally seen as intellectually removed from astrology, including evolutionary psychology. This point bears on a larger contrapuntal lesson that history can teach scientists and the society that supports their efforts. On the one hand, once a grand theory is brought down, its original merits never completely disappear and, indeed, often provide a benchmark for future theories; on the other, there needs to be both motive and opportunity in order to bring down a well-established theory. In the case of astrology, a big part of the motive and opportunity was provided by the early successes of experimental natural philosophy, which led the people we now call 'scientists' to believe that we can control, not merely mediate, the fundamental forces of nature. We might regard current developments in biotechnology as beginning to open up a similar line of attack against the more Darwinian accounts of evolution.

CHAPTER 5

Why Can't Evolutionists Stop Talking About Design?

Theodicy: The Original Science of Design

Contrary to the claims of ID's critics, design is more than a physical pattern that happens to display some interesting mathematical properties, the emergence of which might be explained in any number of ways. Rather, design implies that the pattern constitutes a structure that does, or can, serve a function.[1] However, functions exist only in relation to a system whose parts are differentiated in order to do the things needed to keep the system operative. But who or what is responsible for the system's overall design features, in terms of which one might distinguish functional and dysfunctional patterns of behaviour? The short answer we normally accept in science is that responsibility is in the eye of the beholder. In physics, one speaks of the observer's 'frame of reference' as being responsible for determining an object's position and movement. It is a conventionally chosen 'fixed point', in relation to which all other objects appear to move. After Newton and before Einstein, physicists generally believed that there was only one truly fixed point, namely God's, beyond the many conventional fixed points that we might adopt for the purposes of measurement or calculation.

In the 17th century, theologians turned God's point of view, the design standpoint, into a science in its own right: 'theodicy', which literally means 'divine justice'. From the start, theodicy conceptualised divine justice as a problem of balancing out competing demands or opposing forces, what economists more generally call 'optimisation'. After all, notwithstanding God's supreme powers, creation is always presented as the ultimate struggle of mind over matter – an optimisation problem. In both accounts of creation in Genesis, God faced material resistance to the imposition of his will. Creation never happens all at once. In Genesis 1, God is portrayed as managing a chaotic situation by partitioning his task in spatial and temporal terms, which his creatures then inherit as the framework for orienting their lives. In Genesis 2, God appears as a sculptor gradually shaping clay to a prototype that he is trying to realise. To be sure, in both cases, there is never any doubt that God will succeed. Indeed, his success appears as the empirical order of reality, the nature of which science then sets out to discover.

At this point, theodicy aims to focus the direction of scientific inquiry further. What end was God trying to achieve by having constructed reality in this way? What cost-benefit trade-offs were made along the way to enable this solution? These may seem like far-fetched questions, even sacrilegious ones. After all, isn't God supposed to be all-powerful? Yes, but to be all-powerful is only to say that one always gets what one wants. It does not specify the *modus operandi*. Thus, it is reasonable to ask how reality would differ, had God made the trade-offs somewhat differently: what would have been gained, and what would have been lost? And, controversially, will, can and should these trade-offs be altered in the future? Here questions of science blend into those of both technology and theology, especially as humans are assumed to partake of God's creative capacity.

A useful entry point to the theodicist's mindset is the argument for God's existence that has most exercised philosophers over the centuries, namely the so-called ontological argument. It goes as follows: We can imagine a being greater than that which we cannot imagine. Such a being, by virtue of its maximum greatness, would not simply exist in our limited imaginations but also in reality. That being is God. Theodicists accept the basic structure of this argument, which nowadays most non-theistic philosophers reject. However, this does not reflect greater logical acuity on the latter's part, but rather a different understanding of what we imagine when imagining 'a being greater than that which we cannot imagine'. Theodicists imagine the God of Genesis, whose infinitely overriding power still must deal with the problem of material resistance. Absent this challenge, it is not clear why God would have been driven to create at all, and if so driven, why creation could not have sprung up full-blown and in a perfect state once and for all. The lack of biblical precedent for these alternative visions of the deity led theodicists to presume, perhaps tacitly to avoid the suspicion of heresy, that God's overriding powers are indefinitely extended human ones, implying an inevitable confrontation with something independent of the deity's own design – the ultimate version of what normally faces humans. The ontological argument is normally construed as attempting to prove God's existence. It is perhaps better seen as trying to explain why in the normal course of human existence we think about God in the first place – namely, because we (or at least monotheists) are inclined to imagine nature as something we could have created ourselves, were our normal powers sufficiently extended.

The original theodicist who appreciated the problem most clearly in these terms was the late-17th-century follower of Descartes, Nicolas Malebranche, who argued that the freedom to act, divine or human, is defined by the multiple means available for realising the same end, each subject to its own distinctive opportunity costs (i.e. paths not taken) and resulting in substantially different outcomes that still incorporate the desired end. After all, freedom means very little, even to God, if all the possible outcomes turn out to be *exactly* the same.[2] Nevertheless, despite the resistance that earthly matter poses to divine form, God operates with maximum efficiency. This means governing the widest range of phenomena by the smallest number of principles, a project that continues to be pursued by cosmologists today as the 'Grand Unified Theory of Everything'.

The dominant image of divine optimisation is the achievement of global equilibrium – a harmonious world-system – by resolving internally contradictory tendencies.[3] In the previous chapter, I noted that this sense of divine creation was already embedded in the original Muslim invention of algebra. But why should solving simultaneous equations be seen as a vehicle for doling out cosmic justice? The relevant sense of 'justice' here is that various acts, each one good in itself but left to its own devices, may combine to produce states of the world that appear evil. Either such evil outcomes need to be forestalled, or where they cannot, they must be turned to good. God's task in the manufacture and maintenance of good is very much like that faced by the US founding fathers who, to recall the language of the Constitution, forged a system of government in which opposing powers – executive, legislative, judiciary – are clearly separated and then all subject to checks and balances, rendering them functioning parts in a common political machine, all along presupposing that the people occupying these roles will act from motives other than purely selfless ones.[4]

Of course, the cosmic optimisation problem posed by theodicy ranges over not only 'natural philosophy' (as physics was known in the 17th and 18th centuries) but also 'natural history' (aka biology). In principle, everything is grist for the cosmic mill. Even death and destruction must be seen as part of the most economical strategy for realising the divine plan. But as we are also part of the divine plan, are we not entitled to bring creation to a more satisfying sense of completion? Perhaps, though this was a topic of considerable controversy. In any case, theodicy was the natural outgrowth of a rationalist conception of theology that takes seriously the rootedness of human action in divine action. An apt naturalistic comparison is today's debates relating to the Earth as a unified ecosystem, which take seriously the role and impact of humans as part of nature. Do we provide nature with a sense of direction and even completion, or are we a blight on the planet that is best minimised if not altogether eliminated?

In biology, the design standpoint is captured by the idea of the 'selection environment', which is neither so rigidly nor so quantitatively defined as 'frame of reference' in physics. Biologists assume that there is ultimately one selection environment – the one that has actually operated on Earth – but that its character is established only once certain organisms, organs or cells are 'observed' (or inferred from the historical record) to have been selected.[5] How, then, might one characterise this selection environment? Consider the following three possibilities.

First, imagine Earth as one big environment that humans will eventually populate in its entirety, having superseded all other species. This was the vision advanced by the original theorist of evolution, Jean-Baptiste Lamarck, who secularised the Abrahamic imperative to radically transform the Earth through technology, thereby realising humanity's birthright as creatures made *in imago dei*. From this standpoint, natural selection is the prototype for a superior version of artificial selection by which we shall finally bring order to the planet. Despite their professed Darwinism, the Christian undertow of many leading population geneticists in the early 20th century, including Fisher, Wright and Dobzhansky, encouraged this eugenicist reading of the selection environment. Without speaking the language of eugenics, Teilhard de Chardin probably offered the most self-conscious articulation of this vision.

Second, suppose the Earth is seen as many localised selection environments, as modern evolutionary theory tends to do. We may then follow Dawkins in speaking of an 'extended phenotype', whereby organisms remake their environments so as to improve their selective advantage. But this is conceived in relatively limited terms, always with the potential of backlash, as species become 'overadapted' to an environment that later changes significantly beyond their control. Here Darwin's view that natural selection ultimately trumps all efforts at artificial selection rules supreme, and humans are advised to adopt a modest 'precautionary' view to nature. An apt analogy here is the scepticism surrounding a central economic planner – in this case, a pseudo-Lamarckian eugenic selector – in a world of complex overlapping markets.

Finally, consider the selection environment as one's own body – that some cells rather than others are produced in response to physiological changes caused by an invasion of alien microscopic species. Unsurprisingly, the medical scientists most consistently drawn to Darwinism have been immunologists, who routinely face such microbial challenges in the form of epidemics.[6] To be sure, these can be met, but there are social and technical limits, not to mention side effects. It is no accident that immunologists have been divided over just how much vaccines can or should reverse the inevitable micro-conflicts for survival that transpire in the 'inner environment' that constitutes our bodies. The Nazis easily mobilised devotees of 'racial hygiene' in German medical faculties, who had accepted the idea that the drive towards mass vaccination was a reactionary anthropocentric response to an aspect of natural selection that kept the human population at a sustainable level. They treated microbe-driven epidemics as natural selection's quick way of reducing our impending threat to other creatures. To the racial hygienist – even pre-Hitler – the term 'disease' was prejudicial in its suggestion that microbes had no positive role in the global ecology other than the discomfort they cause to humans. The Nazis merely encouraged the extension of this line of thought so that humans might learn from the microbes to act as agents to cull their own populations.[7]

But couldn't all these value judgements about design be put aside if we simply replaced words like 'adaptive', 'functional' and 'fit' with 'marginally higher survival rates'? I am afraid not. Those three design-based words presuppose a comparison with other possible organisms that might have taken the successful organism's place, but happened to fail to meet the demands set by a specific selection environment. To specify the scale and scope of this 'selection environment', and the relevant alternative organisms to the surviving ones, requires a design perspective. More specifically, the evolutionist cannot escape the role of cosmic problem-solver. The hallmark of problem solving is the need to decide which of several well-defined options best resolves an indeterminate situation for one's purposes. That a situation constitutes a 'problem' implies that one's ends must be achieved within limited means, given one's lack of control over all the relevant intellectual and material resources. This is the optimisation problem that God faced in Genesis.

As the first Genesis account of divine creation suggests, God may work as a modular assembler who treats organisms as what the social science polymath Herbert Simon called 'near-decomposable systems'. Such systems exhibit a complex design, but are constructed to anticipate interruptions prior to completion by the regular saving of work in functionally specific units.[8] Simon, a member of the First Unitarian Church of Pittsburgh from 1955 to his death in 2001, appeared to appreciate the implied theological dimension.[9] The thought experiment he used to illustrate the idea of near-decomposable systems involves, in a nod to William Paley, two watchmakers, Tempus and Hora (named for 'time' in Latin and Greek respectively), only one of whom regularly saves – 'chunks', in the old computer jargon – his work and hence manages to stay in business. In contrast, his imprudent rival envisages the design of each watch in such holistic terms that he has to manufacture each one from scratch after each interruption, and hence finishes many fewer watches.[10]

Among Simon's many contributions was his pioneering work in artificial intelligence from the 1950s to the 1990s, which culminated in a series of computer programs for inducing scientific discovery named after Francis Bacon.[11] Under the circumstances, it becomes easy to grasp his concept of near-decomposable systems, given the dodgy state of computer technology half a century ago, in which large programs were 'chunked' into subroutines as 'backup' just in case the mainframe computer crashed. In that case, once the mainframe recovered, it would be possible to resume from the last completed task in the program. Precedents for this conception can be found in the popular Victorian view of evolution as God's fits-and-starts imposition of design on resistant matter, comparable to the obstacle-ridden tasks of laying down railway tracks or telegraph cables.[12] In more explicitly Darwinian terms, Simon's 'interruptions' can also be understood as the sharp changes in environmental conditions that induce speciation, and probably even more functionally specified genetic clusterings – of limbs, eyes, hearts – that are subsequently incorporated, *au bricolage,* into the construction of new species.[13] In that case, the logic of genomes reflects the divine watchmaker's creation of what cognitive psychologists call an 'external memory store' that can be used and reused to meet design specifications.

Before turning to the ways in which evolution has undermined the richness of design thinking in science, a word must be said about the most famous, if not infamous, personification of theodicy in literature, namely Voltaire's character Dr Pangloss, who never tired of explaining to his ward Candide that every calamity they encountered demonstrated, yet again, that they lived in 'the best of all possible worlds'. Voltaire makes the design perspective appear ridiculous for rationalising any patently irrational sequence of events. Voltaire himself was less bothered by the irrationality of the world than by false attempts to rationalise it. Indeed, his celebrated defence of the critical power of reason was tied to a basic pessimism about reason's capacity to fathom the ultimate structure of reality. What Voltaire's uncharitable reading of theodicy overlooked was that God's grand optimisation problem may require human action as a crucial part of its solution. Dr Pangloss always addressed reality as a spectator trying to catch up with events over which he held no control. But if humans are created *in imago dei,* then we should address reality as God would; namely, as also facing a grand optimisation problem. In that case, the events we experience are like the unfinished matter that God faced at the initial creation. From this perspective, the sort of calamities that befell Candide and Pangloss provide opportunities for humanity to transform the situation to its own advantage. They are invitations to participate in the divine. We are asked not merely to see nature as rational but to *make* it rational.

Making Evolution Safe for Intelligent Design: Gassendi's Legacy

The idea of evolution has been expressed in many different times and places. It is a common feature of most of the great cosmologies, not least the 'karmic' religions of India and China. However, before the modern era, evolution was rarely associated with the idea that reality is constituted as a 'universe' susceptible to human comprehension and control. On the contrary, cultures receptive to evolution have tended towards irrationalism and 'pluriversalism', or what is nowadays called the 'many-worlds thesis', according to which there is no ultimate reality, let alone one that uniquely merits sustained scientific inquiry.[14] Evolutionists through the ages have taken to heart the idea that forms of life are simply temporary arrangements of matter in motion that pass in and out of existence with a certain regularity, if not periodicity, the significance of which depends on the observer's frame of reference. Nature may not even possess an origin or plan, let alone a distinct plan with a unique origin – just the ceaseless rhythm of change. From this point of view, nature is something one adapts to and copes with, rather than fathoms or overcomes. Under the circumstances, nature becomes more an ethical than an epistemological category – that which disciplines us, not that which we discipline.

It is true that the Ancient Greek and Roman schools associated with an evolutionary cosmology, atomism and Epicureanism, advanced metaphysical positions that were congenial to the modern scientific world view. But that happened with no help from the ancients themselves, who thought about science as a form of medical practice – what Martha Nussbaum nowadays dignifies with the name 'therapy' – that tried to persuade patients to become just that; that is, passive recipients of natural forces that would ultimately master them. Wisdom was associated with the release from suffering that accompanied the abandonment of impossible desires for mastery. It was only once atomism and Epicureanism were embedded in a universalist cosmology subject to intelligent design that they contributed to the organised resistance against nature that has been characteristic of modern science. This cosmology derived from the biblical religions, in which the deity, in whose image humans are uniquely created, is presented as engaging in an ongoing but ultimately successful struggle against nature to realise his intentions. This point animated Darwin's great public defender, T.H. Huxley, who spent most of his career promoting the aggressively death-defying, increasingly high-tech approach to medicine that is perhaps the clearest secular descendant of that biblical attitude, which at least for now we continue to take for granted.

Huxley perceptively argued that the logic of scientific progress required that God-intoxication precede detoxification, hence Newton's historical priority to Darwin: the human condition would have remained squalid had they lived in reverse order. In other words, humanity had first to believe in its own privileged grasp of reality in order to justify the sustained effort associated with the pursuit of knowledge and its worldly applications.

Newton's twenty-year theologically driven endeavour towards *Principia Mathematica* epitomised this confidence. However, Darwin's own twenty-year intensive study of nature, initially spurred by a similar confidence, reached the ironic conclusion that we are not so special, after all, differing from the rest of nature at most by degree, not kind. The great puzzle that Huxley posed to the 20th century – and still poses to us in the 21st century – is how to motivate Newton-sized scientific ambitions in the wake of Darwin's deflated view of humanity.[15]

The person who was probably most responsible for converting the traditionally science-stopping evolutionary philosophies of atomism and Epicureanism into science-starters was the maverick French priest, Pierre Gassendi, whose posthumously published *Syntagma Philosophicum* (1658) influenced the likes of Hobbes, Boyle and Newton. Gassendi is best known today as the teacher of Cyrano de Bergerac, who translated the priest's insights into romance and science fiction. For his own part, Gassendi was a systematic theorist who normalised a potentially subversive viewpoint – in this case, one that had already become widespread in the world of Renaissance letters but not among more sober academic minds. If Heidegger made the already popular Nietzsche safe for professional philosophers in the middle of the 20th century, Gassendi had done something similar three centuries earlier for Rabelais and Montaigne, who were widely suspected of paganism for taking atomism and Epicureanism seriously. Both Heidegger and Gassendi functioned as intellectual vaccines that inoculated the literate against more virulent strains of a foreign agent that could have undermined Christendom altogether. In light of Gassendi's work, it became possible to speak openly about atomism and Epicureanism as elements of a Christian world view without appearing ironic or self-contradictory.

In this respect, Gassendi is owed a debt of gratitude by today's 'theistic evolutionists' who would embed a chance-based conception of evolution by natural selection in an overall rational universe. After all, the ancient proto-evolutionists concluded that life is ultimately pointless, and that peace of mind is maintained in the face of our transient existence by minimising pain in the brief time our bits hold together enough to allow for sentience and continuity of consciousness. In practice, this was usually a counsel to minimise effort altogether: the less staked now, the less lost later. Under the circumstances, it was hard to see why anyone inclined to such a world view would invest the time and energy required for technical scientific work, much of which – even by scientists' own accounts – ultimately confounds one's expectations. If there is no larger project to which even one's own failed efforts might be reasonably seen as contributing, then years of drudgery in a laboratory – or, for that matter, poring over ancient texts in an archive – would seem to be wasted. Perhaps some strategically targeted efforts could be justified in the case of medical research, but only if they genuinely aim to minimise pain, and not simply extend its duration, albeit in an attenuated form. It is not by accident that atomists and Epicureans were forthright defenders of suicide and euthanasia, while extending a friendly hand to animals. A Neo-Darwinian variant of the same combination of positions is held by Peter Singer today.

Nevertheless, a problem that haunted Gassendi's attempt to mitigate the nihilistic strains in ancient atomism and Epicureanism goes to the heart of ID. In response to his great contemporary and rival René Descartes, Gassendi ventured that human psychology is not especially well-designed to receive the truth, given our susceptibility to what most immediately attracts the senses. This is a version of the problem that had faced Tertullian in the early days of Christianity. However, Gassendi did not share Descartes' optimism that rational self-discipline informed by Christian principles would enable us to comprehend the divine plan. Instead he concluded that God designed us in such a way that his nature would remain forever elusive, rendering the palpable imperfections of the world-system largely inexplicable.[16] Gassendi did not accept Malebranche's suggestion that a perfect creation need not – and perhaps could not – be perfect in all its parts.

In the short term, Gassendi's plea for humility helped to shore up his Catholic credentials in a time when authorities were keen to identify Socinian heretics who would blur, if not erase, any qualitative difference between human and divine powers. But in the long term, it sowed more general doubts about the reality of design in nature that would be later picked up by Voltaire, Hume and, finally, Darwin. Indeed, thanks to Gassendi, future generations came to see imperfections in nature as a *prima facie* argument against the existence of a divine creator. Nowadays evolutionists give popular expression to such doubts by sarcastically remarking that the jerry-rigged character of organisms, not least ourselves, ill suits a divine creator: a perfect God would have worked from blueprints rather than bricolage. The argument presupposes that a truly competent God would have created each species fully formed, and that the purpose of each species in nature could be understood simply by studying how it functions in its normal environments. More profoundly, it presupposes that our entitlement to comprehend the divine plan could be redeemed with minimal mental effort.

One might think that, given the burden of original sin, it is unlikely that the path to comprehension would be so straightforward. Yet, as it turns out, William Paley was one influential theologian who upheld Gassendi's gold standard of God's perfect creation at the dawn of the 19th century, and not surprisingly Darwin targeted him. Consider the paragraph Darwin wrote just before he issued the challenge that seduced Behe into proposing the principle of irreducible complexity:

> It is scarcely possible to avoid comparing the eye to a telescope. We know that this instrument has been perfected by the long-continued efforts of the highest human intellects; and we naturally infer that the eye has been formed by a somewhat analogous process. But may not this inference be presumptuous? Have we any right to assume that the Creator works by intellectual powers like those of man? If we must compare the eye to an optical instrument, we ought in imagination to take a thick layer of transparent tissue, with a nerve sensitive to light beneath, and then suppose every part of this layer to be continually changing slowly in density, so as to separate into layers of different densities and thicknesses, placed at different distances from each other, and with the surfaces of each layer slowly

> changing in form. Further we must suppose that there is a power always intently watching each slight accidental alteration in the transparent layers; and carefully selecting each alteration which, under varied circumstances, may in any way, or in any degree, tend to produce a distincter image. We must suppose each new state of the instrument to be multiplied by the million; and each to be preserved till a better be produced, and then the old ones to be destroyed. In living bodies, variation will cause the slight alterations, generation will multiply them almost infinitely, and natural selection will pick out with unerring skill each improvement. Let this process go on for millions on millions of years; and during each year on millions of individuals of many kinds; and may we not believe that a living optical instrument might thus be formed as superior to one of glass, as the works of the Creator are to those of man?[17]

This is probably the paragraph of Darwin's that is most often cited to distinguish an evolutionary from a design-based explanation of the same phenomenon. However, the unprejudiced reader might reasonably wonder whether it really does any such thing, given the degree of God-like agency attributed to something called 'natural selection'. Here, Darwin reveals more about his rather strict views on how God must act in order to merit our allegiance than about the workings of evolution as such. Darwin portrays his theistic opponents as believing in a God who created the eye in a single unified act, which he then attacks, interestingly, on *both* religious and scientific grounds. Since Darwin doubted that artificial selection could match the feats of natural selection, he resisted any hint that God might be an amplified version of a genius inventor who created the eye in the manner of the telescope.

Here we should note that in Darwin's day, the invention of the telescope was commonly attributed to Galileo's genius. But nowadays Galileo is cast as only one significant developer of an artefact that previously existed as a toy, which many after him refined as its basis in optics and applications in astronomy became clearer. In other words, our understanding of the actual history of the telescope has come to conform to the hypothetical narrative that Darwin imagined for the eye. This point has not escaped the notice of recent historians of technology, for whom Darwin's extended metaphor for natural selection reads well as a *literal* account of the collective enterprise involved in the intelligent (human) design of artefacts.[18] It involves the distribution of intellectual and material resources over time in order to try out different solutions to cognate problems, the results of which are then drawn upon and supplemented by the next generation of inventors.

Moreover, *Origin of Species* was largely not read in the theologically and scientifically deflationary terms that its author intended. Darwin's tendency to blur empirical probability and logical possibility – or, more brutally, fact and fiction – allowed readers to infer a clear image of the eye's design by evolutionary means. In this regard, they would have been familiar with Whewell's popularisation of Kant, which pushed the meaning of 'possibility' closer to such psychological terms as 'conceivability' or 'plausibility'. As beings created *in imago dei*, if God can do it, then we should be at least able to imagine it, if not do it as well. Thus, Darwin's ability to imagine in some detail the process by which an organ like the eye, which seems so perfectly fit for purpose, could have been the product of small incremental changes over a very long period was tantamount to a simulation of how God could have worked *through* natural selection. In that case, all Darwin showed was that we should not be too hasty in likening God's sense of 'simple' design to our own, but nothing he said precluded our ultimate comprehension of the divine plan.

Darwin wrote at a time when mechanical models of subtle physical entities like the 'luminiferous aether' – the medium postulated to explain the peculiarities of the transmission of light – were treated as the next best thing to proofs of their existence. Indeed, the 19th century is now regarded as the first great era of scientific modelling. Its champions included an array of scientists who would have seen natural science incorporated into natural theology; not only Whewell but also James Clerk Maxwell, the great unifier of light and electromagnetism, as well as the computer's original theorist and holder of Newton's Cambridge chair in Darwin's day, Charles Babbage. Babbage proposed an ID argument whereby natural history amounts to the execution of an open-ended computer program designed by God.[19] All of this changed in the first decades of the 20th century, when relativity theory and quantum mechanics ranged over aspects of reality that defied our unadorned senses. Philosophers of science then began to suggest that the traditional currency of intelligibility – a mental image or a concrete model – was neither necessary nor sufficient to demonstrate the workings of physical phenomena. A set of mathematical equations tied to experiments that produced reliable outcomes would do the trick, even if we could not assimilate the outcomes into our ordinary modes of perception.[20]

With the advent of computer simulations, however, the importance of nature's intelligibility received a new lease of life, as it then became possible to provide a concrete model of abstract representation. Instead of constructing an elaborate machine to demonstrate the plausibility of light's peculiar movements through an aether, we now resort to multi-dimensional computer graphics to represent changes in successive generations of organisms over a hypothesised evolutionary history.[21] Moreover, given the increasing expense and 'real world' complexity of first-order scientific research, scientists have resorted to benchmarking scientific validity to mechanically executed imaginative processes – that is, simulations are taken as 'virtual realisations'. Thus, we find ourselves in the midst of the second great era of scientific modelling.[22]

Some philosophers and scientists have betrayed a faith in this procedure befitting a 19th-century ID theorist – sometimes, ironically, under the Neo-Darwinist banner. Consider the *Kitzmiller* testimony of Robert Pennock, a philosopher who testified for the prosecution but spends most of his time with scientists who spend most of *their* time staring at computer screens. Pennock zealously claimed that a computer program called 'Avida' – which to the naked eye does little more than generate computer viruses within certain parameters – 'instantiates' evolution by natural selection.[23] In metaphysics, the word 'instantiates' refers to the individual embodiment of a common idea. It follows that Pennock believes that 'natural selection' as demonstrated through Avida is the same as that (purportedly) exhibited in the Earth's natural history. This is much stronger than saying that the computer program 'simulates' evolution, which would merely imply that it captures some but not all of the relevant biological processes. In any case, Pennock has moved way beyond Darwin's own evidence base. As it turns out, Pennock's claim went unscrutinised by the defence lawyers and was taken as gospel by the presiding judge. But what could justify Pennock's slide from *simulation* to *instantiation*? Setting aside the possibility of sheer zealotry or sloppiness on his part, two answers suggest themselves. On the one hand, 'evolution by natural selection' has been somehow promoted to a universal law of nature – well beyond Darwin's original, and still controversial, principle for explaining life on Earth. If so, when did that happen? On the other hand, Pennock may be implicitly, perhaps even unwittingly, trading on our presumptive godlike powers to produce outcomes that to someone not involved in their creation do not appear to have resulted from intelligent design.

That the subtleties of divine optimisation – the stuff of theodicy – are rarely appreciated today, even among the religious, is a testimony to how far theology has retreated from a robust design perspective. This reflects the overriding influence of Paley's argument for God's existence, which left the impression that design-based arguments imply a complacent creator whose handiwork can be understood simply upon inspection and admired by a grateful but passive humanity. However, the strongest arguments for design have placed the free will of both God and humans at their centre. Interestingly, a new chapter in the history of theodicy has recently been opened at the online bookseller Amazon.com. Michael Behe and his fellow Catholic scientists, Kenneth Miller and Francisco Ayala, have replayed (unwittingly, I suppose) the 17th-century debate between Malebranche and Gassendi on the blog page for Behe's latest book, *The Edge of Evolution*.[24] All three agree that that there is an 'edge of evolution'. In other words, physical reality cannot be completely explained as some combination of chance and necessity but rather possesses an 'anthropic' character, such that were its parameters slightly different we would not be around to discover them. This idea helps to motivate the theism shared by Behe, Miller and Ayala.

The outstanding question is: when does God 'let go' from micromanaging creation to allow natural selection to take its course in producing the organisms that constitute the history of life on Earth? For his part, Behe holds that God remains in direct control for longer than theistic evolutionists like Miller and Ayala allow. Consequently, Behe controversially attributes to God all the suboptimal outcomes associated with extinction, maladaptation and other natural disasters that can befall species, including our own. Of course, God does not desire these mishaps, or 'evils', yet they are nevertheless unavoidable consequences of the divine plan's realisation. Indeed, to enable a robust sense of free will that would allow humans to develop their divine capacities, the world has had to be created as a dangerous place – that is, where something is genuinely at stake in choosing one way rather than another. Miller and Ayala recoil from Behe's Malebranchian line of reasoning, preferring, Gassendi-style, to imagine that the apparent imperfections of natural selection's *modus operandi* simply reflect an absolute limit in our comprehension of God's work.

In the next chapter, we turn to a scientifically productive precedent for ID's quest for the moment in nature that divides designed from chance-based processes. The case is epitomised by Warren Weaver's career as both pioneer information theorist and godfather of modern molecular biology. For Weaver, the relevant edge bounded 'uncertainty' rather than 'evolution'. Weaver wanted to overcome the apparent limits set by quantum mechanics to determinate human intervention. His proposed solution was to locate the smallest unit of matter that retained its functional properties – what the medieval alchemists called *minima materia* – that could then be manipulated for humanly desirable ends. In the past decade, this quest has resurfaced as an enthusiasm for 'nanotechnology', since a billionth of a metre seems to define the limits of efficacious human intervention in nature.[25] This trend, supported by the leading scientific funding agencies throughout the world today, provides a sense of what the research trajectory of science would look like with ID at the centre of its concerns.

CHAPTER 6

Why is Intelligent Design Unlikely to Go Away?

The Persistence of Design Language

It is striking that evolutionists cannot seem to distance themselves from Darwin's origins as a failed ID theorist. Thus, today they routinely equivocate over whether design in nature is real or only apparent. Evolutionists clearly wish to retain many aspects of the idea of design: words like 'adaptation', 'function' and 'fitness' suggest that organisms appear designed for certain environments, even if they were not specially created with that purpose in mind. Regardless of how the organisms originated, as a matter of fact they can perform in ways that increase their survival rates. This subtle distinction, between how an organism came to be as it is and why it continues to be as it is for as long as it does, leads some evolutionists to freely admit the illusoriness of design in nature. But most are more circumspect.

In any case, three things are clear. First, design-based language shows no signs of disappearing from either the popular or the technical literature in biology. Second, those like Dawkins who wish to dismiss the persistence of design-based language as 'mere metaphor' have never presented a convincing account of the cognitive role of figurative language in science to justify that trivialising conclusion. Third, the history of science across many different fields reveals a tendency for metaphors to graduate to the status of models, and ultimately empirically discernible mechanisms. Indeed, design-based language has proliferated in recent evolutionary explanations, which increasingly rely on computer simulations to assess the 'optimality' of alternative developmental pathways, be they understood at the level of a population, a single organism or a biochemical process.

Consider the following quote, which is taken from the closing paragraphs of an article purporting to provide an evolutionary account of the Krebs cycle, the biochemical process by which food is metabolised into energy:

> The Krebs cycle has been frequently quoted as a key problem in the evolution of living cells, hard to explain by Darwin's natural selection: How could natural selection explain the building of a complicated structure *in toto*, when the intermediate stages have no obvious fitness functionality? This looks, in principle, similar to *the eye problem*, as in 'What is the use of half an eye?' However, our analysis demonstrates that this case is quite different. The eye evolved because the intermediary stages were also functional *as eyes*, and, thus *the same target of fitness* was operating during the complete evolution.

> In the Krebs cycle problem the intermediary stages were also useful, but for different purposes, and, therefore, its complete design was a very clear case of opportunism. The building of the eye was really a creative process in order to make a new thing specifically, but the Krebs cycle was built through the process that Jacob called 'evolution by molecular tinkering', stating that evolution does not produce novelties from scratch: It works on what already exists. The most novel result of our analysis is seeing how, with minimal new material, evolution created the most important pathway of metabolism, achieving the best chemically possible design. In this case, a chemical engineer who was looking for the best design of the process could not have found a better design than the cycle which works in living cells.[1]

Kenneth Miller cites this article as a point-blank refutation of ID, or at least its claim that the cycle is 'irreducibly complex', in the phrase biochemist Michael Behe coined for biological phenomena whose parts are so well integrated that it is highly unlikely they could have come about by chance-based Neo-Darwinian processes.[2] But should conclusions expressed in this design-heavy language be allowed to pass as *refutations* of ID?

Aside from a couple of perfunctory references to the works of Richard Dawkins, the authors do not attempt to inform the concept of evolution with the sort of distinctively Darwinian content provided by evidence from the Earth's natural history. Moreover, not only does 'evolution' function as the active – even 'creative' – subject of a sentence, a grammatical role that in the 19th century would have been assigned to God, but the authors go so far as to propose two contrasting design-based strategies as hypotheses to explain the Krebs cycle that are little more than secular theodicies: evolution as either *optimising engineer*, as in the case of the eye, or *opportunistic tinkerer*, the one they prefer for the Krebs cycle. It would seem that the only design-based position that they have *not* co-opted for evolution is the specific version of ID that Behe advocates in this case, namely that it was the product of a single made-for-purpose creative act without intermediates or precedents in nature. The authors are entitled to push this narrow interpretation of ID only because Behe himself does. But the result is as intellectually satisfying as judging the merits of the Neo-Darwinian synthesis entirely on the basis of Richard Dawkins' extreme formulation of the 'selfish gene' hypothesis.

An important strategic problem facing ID defenders is exactly what to make of the considerable, possibly even increasing, overlap between the language of design that they and their evolutionary opponents use. The existence of such overlap would seem to suggest that the two sides differ more at the level of overall research orientation – what Karl Popper called 'metaphysical research programmes' – than of testable scientific claims issued from the laboratory bench and recorded in peer-reviewed journal articles. This is reflected in the different phenomena with which both believe 'the facts of life' need to be rendered 'consilient', another word coined by Whewell, this time to describe Newton's feat of unifying findings from a variety of disciplines under a set of simple laws.

Within a broad definition of the 'scientific community' (that is, knowledge workers whose expertise is drawn mainly from mathematics or the natural sciences), ID derives its greatest support from fields peripheral to Darwin's original concerns. These include the branches of biology closest to chemistry and physics, as well as engineering – including software engineering – and parts of medicine. In contrast, evolution's heartland is to be found among the historically field-based disciplines in which Darwin himself would feel most comfortable today: zoology, botany and palaeontology. Genetics is a battleground common to both. Yet, truth be told, the emergence of the Neo-Darwinian synthesis in the 20th century has largely amounted to the displacement of Darwin's own competences by people possessing much the same training and sensibility as those now inclined to support ID. It has involved a retreat from the field to the lab, from the study of organisms and populations as they appear to the naked eye to the study of biological phenomena well above *and* below that observational horizon: on the one hand, computer simulations of intergenerational change among the various species living in a common habitat and, on the other, microscopic experimental interventions into the molecular make-up of an individual organism.

Indeed, had the history of biology proceeded as Darwin envisaged, language suggestive of an intelligent designer should have diminished with further research. For example, he himself believed that what modern biologists now a call a 'cell' was an indeterminate mass of protoplasm.[3] To be sure, the organisation of the cell was only beginning to be understood in Darwin's day, though his pessimism was not shared by, say, Gregor Mendel. Moreover, Darwin's understanding of natural selection as radically superior to artificial selection suggested that the nature of life would be forever alien to our species-specific modes of understanding. The last thing he imagined was that the mechanisms of heredity could be rendered in terms as humanly tractable as a genetic *code*, which implies the communication of information – but from whom, and why to us? It comes as little surprise, then, that ID supporters have been attracted to information theory.

Life as a Literal Code: The Quest for Biophysics

ID operates with an anthropomorphic, even literal, sense of intelligence that is indebted to the Abrahamic idea of humans as created *in imago dei*. In that sense, ID supporters remain true to the etymology of 'intelligence', which derives from the Latin for 'understand'. Something possesses 'intelligence' if it can be understood, which is to say if *we* can understand it. The idea is ultimately sociological: something is intelligible only if it involves a meeting of minds.[4] This sense of intelligence is embedded in what is nowadays called 'information theory', which was originally formulated to account for the telegraphic transmission of 'intelligence' in sensitive diplomatic missions.[5] The defining feature of these situations is that an important message specifically addressed to the receiver is conducted through a medium that potentially distorts the message. Thus, there is considerable room for error, but probably *not* because the sender him- or herself is malign. As in humanity's dealings with nature, the biggest threat to successful communication is confusion generated by one's own ignorance, which often takes the form of mistaking half-truths for whole truths.

During the Cold War, cyberneticians formalised this threat as the 'positive feedback loop': that is, an effective activity that turns disastrous when pursued indefinitely.

The theological implications of this concept were never far from the mind of the chief theorist and populariser of cybernetics, the mathematician Norbert Wiener, a product of a Unitarian Christian upbringing. Indeed, he sounded a note of cautious optimism, once he concluded that the devil had no power other than the opportunity provided to him by positive feedback:

> As to the nature of the devil, we have an aphorism of Einstein's, which … is really a statement concerning the foundations of the scientific method. 'The Lord is subtle, but he isn't simply mean.' Here the word 'Lord' is used to describe those forces in nature which include what we have attributed to his very humble servant, The Devil, and Einstein means to say that these forces do not bluff. Perhaps this devil is not far in meaning from Mephistopheles. When Faust asked Mephistopheles what he was, Mephistopheles replied: 'A part of that force which always seeks evil and always does good.' In other words, the devil is not unlimited in his ability to deceive, and the scientist who looks for a positive force determined to confuse us in the universe which he is investigating is wasting his time. Nature offers resistance to decoding, but it does not show ingenuity in finding new and undecipherable methods for jamming our communication with the outer world.[6]

Wiener proceeds to contrast science with politics, which does involve active deception, but which is itself based on mutual ignorance, suspicion and failed communication. The famous paradox of game theory, 'The Prisoner's Dilemma', epitomised this conception of politics. The great promise of cybernetics was that the identification of nature's control mechanisms would reduce, if not end, antagonism among humans, who could then join in common cause to exercise dominion over nature in an equitable fashion.

While such a vision might now appear incredibly naive against the backdrop of the Cold War, it was widely shared by other scientific theorists of a Unitarian bent. Not least among these was Warren Weaver, director of the Natural Sciences Division at the Rockefeller Foundation, nowadays known mainly as co-author, with Claude Shannon, of *The Mathematical Theory of Communication* (1949), the landmark work that cast the transaction of information in terms of the purely physical parameters provided by the laws of thermodynamics. It was here that information first came to be presented as the reduction of entropy, or 'negentropy', where entropy is anything that generates uncertainty about what is being communicated. Weaver regarded information theory and, as we shall see below, molecular biology as a two-pronged attack on the mystifying claims of the indeterminacy of reality at the quantum level, which threatened to undermine belief in a divinely ordered universe. For Weaver, Heisenberg's Uncertainty Principle – which denies that a subatomic particle's momentum and position can be measured at the same time – reflected a state of human ignorance remediable through keener measuring instruments and/or deeper theoretical reflection, not a fundamental limit on our grasp of physical reality. He did not accept the widespread view that Heisenberg had demonstrated that the observer participates in constituting quantum phenomena in a way that precludes any determinate understanding of what is constituted. That would be to suggest that we remain forever embedded in nature with no prospect of transcendence, by which we would come to acquire the divine frame of reference.

Shannon and Weaver's model of communication was the noise-filled medium of walkie-talkies that had been used in the Second World War. The task of their theory was to quantify the signal-to-noise ratio, on the assumption that a very elaborate signal would also be likely to generate considerable noise, thereby undermining the message's informativeness to its receiver. Informativeness is measured by the degree of uncertainty that a message reduces for its recipient; an optimally informative message would leave the recipient with no doubt about what had been conveyed. For those, like Weaver and Wiener, who were possessed of a biblical imagination, information theory's practical grounding in wartime communications could be easily leveraged into the ultimate theological problem: How can we become reconnected with our spiritual mission, given that God has had to communicate through the noise-filled medium of our animal bodies?

However this feat is managed – and Weaver had no doubt that we were up to the challenge – it will always be a risky enterprise. Hasty decisions are bound to be made along the way, which on Weaver's watch included the gross identification of specific bits of genetic material with socially relevant traits and the promotion of nuclear radiation as a vehicle for evolutionary enhancement. But such recklessness did not make fathoming life's mechanics any less feasible, desirable or, indeed, necessary. Weaver shared Wiener's cautious optimism about our capacity to persevere through these excesses in judgement to bring a final order to creation. After he retired from the Rockefeller Foundation, Weaver made this point by observing that religion is already 'perfect' in its vision, and that the challenge of science is to muster the requisite intellect and will to realise that vision, all of which sounds like an updated version of Joseph Priestley's Unitarian science policy for establishing a 'heaven on earth'.[7]

Weaver not only gave the field of molecular biology its name, but also funded the retooling of Cambridge University's Cavendish Laboratory to enable it to launch the molecular revolution. In the decade leading up to the discovery of DNA in 1953, the molecular biology lab was still staffed primarily by people trained in the physical sciences, the traditional focus of Cavendish research. Included among them was Francis Crick, who was working with methods that had been developed for use in physics and chemistry, such as X-ray crystallography and digital computing. These methods helped to capture and store large amounts of data about the structure of complex molecules. At the same time Cambridge itself appeared haunted by the ghost of its alumnus, Charles Darwin. At least, that would make a neat supernatural explanation for the university's resistance to pioneering work such as Crick's, even after the discovery of DNA. Indeed, by 1962, what is now known as the Laboratory for Molecular Biology was moved two miles off campus to an independent facility funded by the UK's Medical Research Council, leaving the Cambridge biology faculty in the relatively unique position among world-class universities of being dominated by naturalists of the sort that would have made Darwin feel most at home.[8]

The history of biology in the 20th century tends to be written as an account of the role that Darwin's theory of evolution by natural selection increasingly played as an overarching explanation of living phenomena, resulting in its paradigmatic status within the discipline by the late 1940s under the rubric of the 'Neo-Darwinian synthesis'. What makes the synthesis *Neo*-Darwinian is of course its inclusion of Mendelian population genetics as evolution's underlying mechanism. Heredity had eluded Darwin, who believed that an organism's potential was no more than the sum total of the traits exhibited by its ancestors. And while there is nothing historically inaccurate about claiming the triumph of Neo-Darwinism, especially when expressed in these general terms, other stories could be told of biology over the past century that would reduce the forging of the Neo-Darwinian synthesis to a sideshow of largely political and cultural import.

Even now, at the dawn of the 21st century, Darwinism's hold over the public imagination worldwide is strongest when aligned with some kind of creationist sensibility, even if only the belief that God himself is responsible for the process credited to Darwin. At the time of writing, no Nobel Prize in the biological sciences (aka Physiology or Medicine) has been awarded to a Darwinist for work that requires being a Darwinist. To be sure, the Neo-Darwinian narrative is promoted by the scientific establishment, and its ideological function in banning religiously inspired arguments from scientific discourse is transparent. But if we focus on both the general drift of funding for biological research and its multifarious social impacts, they have been more to do with life's generative capacities than the sheer historicity of life. In this context, Darwin is significant as no more – and no less – than a great natural historian, someone who arrayed an unprecedented range of data about life in a theoretically suggestive way. But Darwin's own attempts at causal analysis were so feeble as, in effect, to entrust the matter to others. In terms of the history of physics, Darwin was the last of the medieval encyclopaedists: he was certainly no Newton.

To tell the history of 20th-century biology with Darwin in this diminished role would be akin to the way that, say, Marx and Freud are normally treated in the history of 20th-century social science: not inconsequential but not the dominant paradigm. In this alternative history of biology, Darwin's grounding in field studies and natural history would have been displaced by 1910 as the primary empirical base of the discipline, largely through the laboratory-based work of figures like Hugo de Vries, William Bateson and Thomas Hunt Morgan, all of whom contributed to the rediscovery of Mendelian genetics. The one person whose career perhaps best exemplified the shift from the field to the lab was Theodosius Dobzhansky, who originally trained as a field biologist in Russia but moved to the US to work with Morgan at Columbia University. Dobzhansky inherited the directorship of Morgan's fruit fly lab, from which he wrote the classic *Genetics and the Origin of Species* (1937), the book that properly launched the Neo-Darwinian synthesis.

This alternative history would focus on the schism that emerged within the community of lab-based biologists, starting in the 1930s when the Rockefeller Foundation under Weaver started providing financial incentives for physicists and chemists to migrate to biology, which resulted in the new science of molecular biology.

Weaver's original plan was to ground population genetics more securely in the physical sciences, which would eventually lead to the ultimate natural science, 'biophysics'.[9] Weaver succeeded halfway: molecular biology now dominates biology as a whole, but without eliminating the old population genetics or – more significantly – the mentality informing it. Population geneticists are interested in the propagation of traits across generations of individuals over time, while molecular biologists are concerned with the organic potential of various biochemical combinations in particular individuals.

In terms of the classic biologist's dichotomy of structure vs. function, population geneticists follow an organism's functions, while molecular biologists follow its structures. Thus, when population geneticists caution against eugenicist policies, they are mindful that a given (un)desirable trait might have been caused in so many different ways that attempts to alter the trait's frequency at the level of the trait itself – such as by sterilising everyone with a substandard IQ – are bound to be unreliable. For them, the distribution of traits in a population is a naturally recurrent phenomenon to be observed, not an object of potential manipulation. On the other hand, a molecular biologist addresses the matter the other way round, more like an engineer, wondering what traits are likely to result from a particular biochemical construction embedded in a given organic environment.[10]

The contrast is evident in the wildly negative and positive attitudes towards the prospects of bioengineering displayed by, on the one hand, the risk-averse Richard Lewontin and, on the other, the risk-seeking Walter Gilbert, both professors at Harvard – the one of population genetics, the other of molecular biology. The former writes as if we can do little more than speculate about (physical) causes from known (behavioural) effects, whereas the latter writes as if we can exert sufficient control over the causes to test out their various effects under various conditions.[11] This unresolved awkwardness between the two fields, routinely masked by the popular Neo-Darwinian displays of ideological unity, is well captured by the most authoritative historian of molecular biology, Michel Morange:

> To return to the fundamental problem: what has molecular biology brought to the understanding of the mechanisms of evolution? The answer is short: virtually nothing, for the simple reason that the two disciplines have not interacted. Although both disciplines use the word 'gene', they have not sought to bring their two meanings closer, or even to confront them. For the molecular biologist, a gene is a fragment of DNA that codes for a protein. For a population geneticist, it is a factor transmitted from generation to generation, which by its variations can confer selective advantage (positive or negative) on the individuals carrying it.[12]

The ultimate source of this contrast in the understanding of 'gene' is Erwin Schrödinger's response to the challenge from his younger colleague, Max Delbrück, over the direction that biophysics should take. It is preserved in a series of Schrödinger's inspirational 1943 Dublin lectures, entitled *What is Life?* Both Schrödinger and Delbrück were trained physicists, the latter having worked under the former. But they saw the relationship of the life sciences to the physical sciences rather differently. Much of Schrödinger's later career, like that of Einstein's, was devoted to fending off more influential interpretations of quantum mechanics that would portray the universe as not merely improbable but downright indeterminate. In contrast, Delbrück sided with Niels Bohr, who had abandoned an absolute point of view in physics, having reconciled himself to reality's irreducibly relational character.

In popular histories of science, this is sometimes portrayed as Schrödinger's 'reductionism' versus Delbrück's 'holism', but a more illuminating sense of the difference in perspective has been presented by the late Canadian biophysicist Robert Rosen, who cast the matter in old Newtonian terms: Delbrück's 'force-like' versus Schrödinger's 'mass-like' conception of genes.[13] Whereas Delbrück continued to see humans at the receiving end of natural history, Schrödinger confidently predicted that we could master its workings. Thus, Delbrück identified genes with an organism's manifest traits (the biologist's equivalent of 'expressions of force'), while Schrödinger saw them as simply a subset of all life-producing molecular combinations (i.e. the masses that exert force). To be sure, Schrödinger's message failed to shift the course of cutting-edge physics. Nevertheless his confidence inspired young physicists like Francis Crick, as well as his DNA co-discoverer, the zoologist James D. Watson, to complete the revolution in molecular biology that a similarly minded Warren Weaver had planned.[14]

Schrödinger took the 'physics' part of biophysics rather seriously. Newton's laws had aspired to range over all *possible* mechanical systems, not just the ones we normally encounter. Despite the substantial revisions to

Newton that relativity and quantum theories made at the extremes of physical reality, the universalist aspirations of physics persist, now popularised as the search for a 'Grand Unified Theory of Everything'. Here Schrödinger was especially influenced by his teacher's teacher, Ludwig Boltzmann, who in the final years of the 19th century had successfully unified thermodynamics and mechanics in a single statistical science of energy transfers. Schrödinger accepted Boltzmann's view that physical laws are 'deterministic' in the sense of describing all degrees of freedom, which by the epilogue of *What is Life?* Schrödinger had interpreted as being all the possible points of divine intervention.[15] For example, if God created a world with certain physical properties, he would also be forced to include other properties in order to render that world coherent. But whether God decided to create such a world in the first place was up to him. Schrödinger also accepted Boltzmann's reasoning that, since over time mechanical systems tend to move from a state of order to one of disorder, we must live in one of those very rare worlds that display sufficient order to enable us to take the measurements and do the calculations needed for reliable physical knowledge. Thus, Schrödinger concluded with both Boltzmann and the man who first legitimised statistical thinking in physics, James Clerk Maxwell, that the mechanical system we call our universe enjoys a unique – we now say 'anthropic' – privilege.

Delbrück was much more cautious in his prognosis for biophysics. Despite having researched in areas closer to molecular biology than Schrödinger had, as well as having been instrumental in institutionalising the field in both the US and Germany, he believed that molecular biology should be used to resolve only known heritable traits, effectively subsuming molecular biology under population genetics, itself subject to forces of random mutation and natural selection beyond the strict scope of experimental physical inquiry. Needless to say, under the circumstances, genuine genetic enhancement, let alone the creation of new life forms, would not be part of Delbrück's biophysics. Of course, in the future evolution may throw up new heritable traits, after which their molecular components may come to be known. Indicative of this attitude was the work for which he was awarded a Nobel Prize in 1969 – an experimental demonstration that bacterial resistance to viruses results from random mutation and not organic adaptation. In short, Delbrück did not countenance the idea that molecular biology might provide the basis for a rational exploration of unrealised biochemical combinations for their genetic viability.

Yet, it was just this prospect – that the biophysicist might try, and ultimately succeed, to create life from scratch – that Schrödinger proposed in his Dublin lectures. Moreover, his proposal was made in a spirit easily recognised today: that of testing out various syntactic strings in a 'genetic code' for whether they send a stable biological message – that is, produce a sustainable form of life, which in a theological vein could be interpreted as comporting with the divine plan. As we shall finally see, this idea of life as literally a language also lurks behind biblical literalism.

CHAPTER 7

Who's Afraid of Biblical Literalism?

The Problem of Literalism: Transmitting the Truth in a Faulty Medium

A literal reading of the Bible has done more to help than hurt science over the centuries. This claim is shocking only because we rarely stop to consider what 'literal' means in this context. The multiple senses of 'literal' can be captured by a makeshift 'Hegelian dialectic' – that is, a three-stage argument in which the opposing features of the first two stages cancel each other out, to produce a conclusion that incorporates everything worth preserving in the first two stages without their limitations. Consider, then, the meaning of the following biblical passage, which purports to record what happened when Joshua implored God to impede the sun's passage in order to halt the advance of the enemies of Israel:

> And the sun stood still, and the moon stayed, while the nation took vengeance on its foes ... The sun

> halted in the middle of the sky; not for a whole day
> did it resume its swift course.
>
> (Joshua 10:13)

There are three successively subtler ways of interpreting the passage, all of which can be reasonably called 'literal':

1. It can be read as a straight eyewitness account, as if reported in a daily newspaper.
2. It can be read as an historical document, in which case one would want to know the semantics of the original Hebrew text.
3. It can be read as a piece of constitutionally binding legislation, in which case one would also want to know the intended effect of the text on the original audience.

This series of readings constitutes a Hegelian dialectic because (3) recovers the immediacy of (1) but as filtered through (2). In other words, the third reading presumes that the biblical passage is addressed as much to today's readers as to those over three millennia ago when the event allegedly occurred. However, it does not follow that had the passage been addressed exclusively, or primarily, to the present-day reader, it would have been written as it originally was. In that case, a different event might have been recounted.

In what sense then does (3) remain a literal reading? The answer is that we are addressed in exactly the same sense as the original Israelite audience. In other words, the Bible speaks to us now as it did to those who would have known of Joshua's exploits. Implied here is the idea that the biblical passages are to be understood as literal *acts of communication*, specifically between God and the transhistorical community of potential believers. We are thus positioned as recipients of the divine message, which in turn demands our acknowledgement and (hopefully) assent.

Equally implied is that despite the constancy of the divine message, our physical nature makes us so unreliable as media for its conveyance that we potentially degrade the message through repeated transmission over time and space. As we saw in the previous chapter, making sense of this situation – the cognitive equivalent of evil – preoccupied such founders of modern information theory as Norbert Wiener and Warren Weaver. But it also goes to the heart of modern translation theory, which not surprisingly was originally concerned with the Bible.[1] Starkly put, there is a trade-off between form and content, the terms of which are determined by context: to mean now what the Bible meant when it was written, we would now write differently; and to write now as the Bible was written, we would mean something different. In this respect, the modern genres into which the Bible tends to be neatly fitted – say, as a body of empirical truths, a miscellany of fictional tales, a set of moral precepts and exemplars or some pastiche of these – *all* fail to take the Bible literally.

One person who tried to address this matter straightforwardly was a confidant of Newton's, the theologian and mathematician John Craig. Taking into account the issues just raised, he predicted the second coming of Jesus by calculating when the reliability of our grasp of the Gospels will have become zero: AD 3144. For Craig, this would be the moment when humanity's moral and epistemic orbit is corrected once and for all, comparable to the regular divine corrections of the celestial orbits that Newton postulated in order to balance his calculations. Whatever we now make of Craig's mathematical theology, it attempted to capture an interesting intuition that is rarely registered in modern theories of empirical knowledge: namely, that the quality of empirical evidence naturally deteriorates, especially as more people come to rely on the testimony of observers communicating in their own languages, embedded in their own concerns and increasingly remote from each other in time and space.[2] While officially distancing themselves from a crude image of knowledge as built by the accumulation of solid facts, both philosophers and scientists normally take this view for granted. This is curious. Almost everything else normally decays over time unless actively preserved, which in turn often involves some form of reproduction, if not reinvention. But facts, once established, appear to endure until they are formally overturned – even if the people, theories and methods associated with the fact's establishment have been discarded, discredited or otherwise consigned to oblivion.

Moreover, when facts are overturned, as most inevitably are, their demise is normally blamed on something already present at their birth that had escaped notice over the years, including various errors of measurement, calculation and judgement or a lack of relevant background knowledge that, had it been known, would have altered the original investigator's conclusions. Blame is hardly ever assigned to later investigators who might have misunderstood the original context from which the facts emerged, which led them to assume more similarity of research concerns than was warranted. The very idea that the transmission of knowledge might be an inherently corrupt process is dismissed from the outset. Thus, the radical shift in emphasis in biological research from the field to the laboratory over the past 150 years has not caused biologists to lose faith in the relevance or validity of Darwin's original insights, even though in the end he never believed we would ever fathom natural selection in the way we have.

But were biologists to lose their faith in Darwin, the distinction between science and its history would dissolve, opening the door to historians of science criticising today's scientists with the same authority allowed to their fellow scientists. A modified version of just this situation applies throughout much of the social sciences, which explains the difficulty of advancing credible generalisations: observations made and data gathered about the human condition in times and places distant from our own, however true to their original context, are presumed to have lost something in the translation from them to us, unless comparable empirical phenomena can be replicated, broadly speaking, 'here and now'.

Craig's British Protestant audience took his argument to justify bearing witness for oneself to what others claim to have observed – as opposed to simply relying on the word of priests and theologians who have authorised their own testimony. A qualitative version of Craig's principle had been enshrined in the Royal Society's motto, *Nullius in verba*. It specifically targeted the Jesuits' Counter-Reformation proposal, backed by Aristotle's conception of induction as *epagoge*, the adduction of an exemplary case, that a duly authorised (or 'expert') witness reported more than just his or her experience but also the experience of any rational sentient being.[3] Ironically, as we saw in chapter 1, the Royal Society has now come full circle to take the Jesuit side of the argument. At least, that would make sense of its trumped-up claims for a 'scientific consensus' that are based not on the first-hand experience of the scientific community, but on an agreed chain of command in whose first-hand experience everyone else is expected to trust.

The Jesuit proposal makes no more – but no less – sense than the idea that a social contract forged centuries ago remains binding on successive generations, and that all subsequent legislation is justified on its terms, unless it is formally overthrown. To be sure, in republican polities, this condition is mitigated by regular elections that allow a change at least to the direction in which the constitution is applied and interpreted. But some democrats, notably the US founding father Thomas Jefferson, believed that the social contract ought be periodically renegotiated from scratch, as he put it: 'The tree of liberty must be refreshed from time to time with the blood of patriots and tyrants,' a sentence he wrote *after* the American colonists had secured independence from Britain and ratified their national constitution. Informing this sentiment is Craig's intuition that over successive generations, the likelihood increases that citizens will have lost touch with the original covenant that bound them together. The closest that science today comes to expressing this sensibility is the Popperian falsificationist ethic, in which the burden of proof is regularly placed on a theory to justify its continued endorsement beyond its original evidence base.

The Solution: The Literal Truth is Performed, Not Represented

The special interpretative problems posed by the Bible stem less from its sheer sacredness than from the indefinite multiplicity of its intended audience. How can one document address all of us in the same sense as it did the original Israelites, especially in matters of action? Defining an appropriate sense of 'literalness' in this context is daunting, to be sure, but it is only quantitatively, not qualitatively, different from what legislators and judges face when interpreting foundational legal documents that explicitly bind successive generations living in the same society. Both religious and legal interpretation – often one and the same – were traditionally subsumed under the discipline of 'hermeneutics', whose meaning unfortunately has now lost this specificity, as the term casually ranges over the systematic criticism of any literary or historical text. Yet the stricter sense of 'hermeneutics' highlights the difference between establishing the literal meaning of a text when the reader *is* and *is not* part of the text's intended audience. In the latter case, which is normal for most historical documents, the interpreter's task is more circumscribed. It amounts to eavesdropping on someone else's conversation. The Bible, by contrast, demands not only comprehension but also a personal response amounting to acceptance or rejection of the message received.

It is important to appreciate what is and is not entailed by this approach to biblical literalism. To treat the Bible 'literally' is *not* to take what it says as a fixed doctrine against which to judge the merits of secular claims to knowledge. That would be to use language simply to display authority, precisely the function that led even the pagan Socrates to complain about publicly posted laws in Egypt and the Near East, which instilled in him a regrettable prejudice against writing. Nevertheless, the sort of distinction that Socrates drew between speech and writing is captured in biblical cultures as a difference between the 'performative' and 'representational' aspects of language use. Socrates was clearly a performativist. More generally, Protestants are performativists, while Catholics are representationalists.

The representational use of language stresses the *difference* between the text's source and its receiver. The Bible is read as presenting only some aspects of an overall reality, the fullness of which lies beyond our comprehension. Language is thus used to subordinate, not empower. 'Submission to God' here means sheer obedience to the word of God, or his authorised mouthpieces, regardless of one's own considered judgement. In contrast, a literal understanding of the Bible is performative. It stresses the *similarity* between the text's source and receiver. This allows the book to be read as a script that can be enacted on whatever stage the drama of one's life happens to be played out. 'Submission to God' in this sense entails assuming the role that God assigns to humanity as the being uniquely created in his image. It is tantamount to what an actor does, as he or she tries to inhabit a role by re-enacting the scripted character's thoughts in the context of a particular theatrical production. Thus, as in, say, a production of Shakespeare, what it means 'to live a biblical life' is likely to vary according to time and place, but in each case the legitimacy of the performance will rest on its fidelity to the original script, understood as a communicative transaction.[4]

Biblical literalism owes its performative character to the missing conceptual link between Athens and Jerusalem, what the Greeks called *mimesis*.[5] Today we would say 'method acting', a technique introduced by Constantin Stanislavski in the early 20th century and inspired by Anton Chekhov. It has become a staple of contemporary film courtesy of Lee Strasberg, who trained the likes of Dustin Hoffman, Jack Nicholson, Al Pacino and Robert De Niro. The idea is that one should try to become the person scripted in the dramatic role. Among other things, this means that the filmed lines deviate somewhat from the ones scripted. For only once the actors live the lives of their characters can they determine the best way to convey those lives to an audience, on the basis of which they can then justify altering the original wording. The theological implication is that humans are method actors in terms of the roles scripted for them in the Bible. An exemplary adherence to this principle explains the boldness of *Principia Mathematica*, in which Newton arguably attempted to play the role of the divine creator himself; and what we call 'modern science', associated with the aftermath of the West's Scientific Revolution, is generally best understood as an affirmative response to a literal interpretation of the Bible.[6]

That this claim seems more outrageous than obvious reflects two facts: that multiple interpretations have resulted from this striving for literalism, and that restrictions have been increasingly placed on appeals to the Bible as grounds for legitimising scientific practices. The charter of the iconic scientific body, the Royal Society of London, enshrined this principle in a principled prohibition of matters of religion (and politics, rhetoric, metaphysics) from discussion at its meetings. This was not because its members were anti-religious, let alone friendly to atheists. Quite the contrary: it was because they wanted royal protection from whatever politically charged disputes relating to biblical interpretation might be raging in society at large, which in recent memory had resulted in civil war.

Thus, when one reads the rather blank but positive references to God in the scientific publications of, say, Robert Boyle or Isaac Newton, one should not conclude that they were mouthing theological pieties without conviction. Rather, they were avoiding controversy by refusing to connect their science to a specific understanding of the divine word. In contrast, their contemporary Thomas Hobbes devoted the second half of his magnum opus, *Leviathan*, to biblical interpretation, making it all too clear how the nature of divine action backed up his uncompromising materialism. Needless to say, he was refused membership of the Royal Society – and chased out of England. Like his role model Galileo and his mentor Francis Bacon, Hobbes believed that science and theology refer to the same divinely inspired reality, not simply two segregated realms of human experience. So did Boyle and Newton, of course, but they made the point with much greater tact by refusing to reveal openly how the Bible spoke to them.[7]

When considering the transformations that Christianity has undergone in the modern era, perhaps a more profound tendency than secularisation has been what the historian of philosophy Jerome Schneewind has dubbed the 'devolution of the divine corporation', whereby powers that had been exclusively associated with or authorised by God come to be 'devolved' to humans, either as a whole, through particular societies or in unique individuals.[8] This devolution is roughly traceable through the decline in the divine status of 'natural law' in both politics and science. It is arguably a by-product of the tendency for literal readings of the Bible to support what I discussed in chapters 2 and 3 as a 'univocal' conception of being, whereby God and humans are taken to differ only by degree, not kind. Thus, once the genealogy implied in the Trinitarian idea of God as the 'father' of Jesus who is in turn our 'brother' is taken literally, God and humans become members of the same species who ultimately differ only by order of descent. In that case, only a short logical leap need be made to conclude that the divine plan is simply the plan that humans at their best would have designed for the constitution of reality, had they lived first. At that point, an orthodox Trinitarianism morphs into a heretical Unitarianism. That view comes into its own with the widespread Enlightenment notion, canonised by Kant, that the animal genus *Homo* earns the title of *sapiens* once it becomes released from its 'nonage', or childish dependency, from such would-be earthly surrogates of the divine parent as kings and priests, and exercises autonomy in the moral sphere comparable to God's legislation in the natural sphere. From this point of view, God chastises the suffering Job not because he questions God's motives, but because Job refuses to take matters into his own hands.

Perhaps the best way to encapsulate the spirit of biblical literalism is to contrast two ways of thinking about the theology that underlies the tripartite relationship between language, our minds and the world. For rough-and-ready historical reasons, I shall call these the 'Catholic' and the 'Protestant' ways, though I do not deny that particular Catholics and Protestants have fallen on opposite sides of this divide. I have already characterised the difference in terms of an emphasis on the 'representational' vis-à-vis the 'performative' aspects of language use, the former always incomplete and the latter potentially self-completing.

The Catholic way focuses on language as a parochially human medium for understanding nature. However, nature, as an independent creation of God's, does not require human mediation for its legitimacy. From this standpoint, the various creatures of nature possess a significance that far exceeds what can be captured in words. It follows that anything said about nature is always incomplete. If any form of language can fathom the secrets of nature, it is likely to be the figurative form, as in poetry, which is how the Bible should be read. The multiple interpretations to which such a reading lends itself are in keeping with the inexhaustible character of God's creativity. Under the circumstances, nature becomes fraught with symbolism that is arguably better understood through visual and aural imagery than bare words on a page.

Little surprise, then, that Catholicism has tried harder to foster a receptive climate for faith through an array of colourful icons, sonorous hymns and elaborate rituals – what the Jesuits originally called 'propaganda' – than a diligent reading of the Bible itself. The latter, after all, might encourage the faithful to think through matters of doctrine for themselves, resulting in a proliferation of partial understandings, given language's limited capacity to reveal the entire truth. Of course, this is how Catholics typically regard the Protestant Reformation – an outgrowth of the dissemination of Bibles translated into the spoken languages of Europe. But the Catholic sensibility also carries on through the modern period in what is often portrayed as the humanistic resistance to the 'language of science' (however defined) as the royal road to reality. In other words, the Catholic antipathy to literalism is not limited to the Bible, but extends to any discursive formation that claims a privileged access to the truth. We first encountered this attitude when discussing Giambattista Vico in chapter 2.

Emerging here is an interesting, if somewhat paradoxical, contrast between the Catholic and Protestant positions. The Catholics deny privileged access to the truth, a principle they enforce through the activities of a priesthood allegedly descended from the original apostles of Jesus. For their part, the Protestants affirm the possibility of privileged access to the truth, but only if it is accessible to everyone as children of the same God. Postmodernists can appreciate the Catholics' surface relativism and multimedia sensibility – little wonder that 'the medium is the message' guru Marshall McLuhan was a devout Catholic – but they cannot abide by its subordination to a supernatural goal. Scientists may admire Protestant forthrightness with regard to pursuit of the truth, but recoil from the very personalised form of decision-making ('to accept Jesus into one's life') that is associated with the validation process, as it creates an instinctive resistance to *all* forms of collective human authority including the scientists' own.

Of the two attitudes towards the Bible, the Protestant one is clearly more favourable to literalism, recalling that Judaism, Christianity and Islam are not called 'religions of the book' for nothing. God communicates most directly with humans by the means through which nature itself was created, *logos*: 'the word'. He creates by calling things into existence in the right order. Our own participation in the divine project requires that we too become literate, so as to read the Bible for ourselves and, to the best of our abilities, rethink God's thoughts and re-enact God's deeds. This requires close study of the original text to discern the underlying grammar of divine speech, which provides clues to the logic of God's thoughts, so that we may bring them to their full realisation in our speech and action. Nature is, from this standpoint, simply an alternative expression of this *logos*, one of Francis Bacon's two divinely inspired 'books': knowledge of Hebrew and Greek needs to be supplemented with mathematics and genetics. This point has recently been put with admirable clarity:

> The Watson/Crick revolution has shown us that DNA is all words. Genes are digitally coded text, in a sense too full to be dismissed as an analogy. Like human words they have the power to hurt, and that power is the greater because, given the right conditions, DNA words can dictate with stronger predictability than most human imperatives.[9]

That Richard Dawkins happens to be the author of this quote shows that biologists have not so much refuted theologians as simply assumed their role in a bloodless coup. To be sure, this has occurred offstage from the various highly publicised 'science vs. religion' debates to which Dawkins himself is often party. However, my point is not to contest Dawkins, who I believe understands the state of play exactly right. Indeed, for good measure, I would throw in linguistics and computer science as potential co-workers in the domain towards which Dawkins gestures. But I wonder whether there is now any principled difference between theology and biology, once genetics is taken to constitute – not analogically but literally – the grammar of life. Galileo and Bacon would have been pleased, but not those who continue to promote a 'good fences make good neighbours' policy for relations between science and religion.

Conclusion

Neo-Darwinism and ID face complementary challenges. Neo-Darwinism needs to justify the continued pursuit of science, given the diminished cosmic status that the theory accords to our species and the ecologically destabilising consequences of the science that we have increasingly pursued. For its part, ID needs to adopt a consistently progressive stance towards the pursuit of science, as befits creatures designed *in imago dei* to master nature. If this dual challenge seems disorientating, that is only because, on the one hand, Neo-Darwinists continue to dine out on ID-based reasons for esteeming science as the signature project of human privilege, while on the other, ID theorists have yet to take the full measure of the literal force of our biblical entitlement, which requires embracing, however tentatively, science's Faustian dimension.

Neo-Darwinism continues to be promoted as a 'progressive' approach to science, even though it would seem to demonstrate the self-limiting character of science, while ID theorists remain coy about the ultimate causes of our scientific understanding, even though clearly they believe that science's enormous success reflects our intimacy with the divine. Thus, the difference between

Neo-Darwinism and ID does not boil down to a simple choice of 'science vs. religion' or 'science vs. non-science', let alone 'good vs. evil'. The two research agendas have historically overlapped in content and have produced or inspired both good and bad consequences. In this respect, advocates on both sides could help to nuance the discussion by acquainting themselves with 'dirty hands' histories of their own movements. Two exemplary ones are, respectively, Richard Weikart's *From Darwin to Hitler* and David Noble's *The Religion of Technology*. Not surprisingly, both books have generated many hostile responses for the inconvenient truths they reveal.

But beyond this, and by way of conclusion, let me outline a strategy to take forward the ID position since, unlike evolution, it has yet to realise its full potential in the public debate over science policy.

The legal and political failures of creationism and ID to date should not be a source of undue embarrassment. They simply show that if one is excluded on ideological grounds from the resources normally needed to amass and promote research teams, including access to jobs and journals, then a politically stable democracy leaves one with recourse to nothing but the courts and the media. Indeed, if the American Civil Liberties Union were truly confident that ID is such rogue science, it would declare its willingness to provide legal assistance *pro bono* to parties wishing to exclude it from public school science classrooms. In that case, potential ID supporters would not feel intimidated by the prospect of having to pay exorbitant legal fees if they lost the case, as in *Kitzmiller*, which bankrupted the Dover school board.

Much of the discontent generated by the prospect of creationism, or even ID, being introduced into science classes rests on two confusions that evolutionists tend to promote. The first is a failure to distinguish between attempts to remove evolution from the curriculum and attempts to add some form of creationism or ID. The spirits of the two proposals are rather different. Calls for the removal of evolution tend to object to the theory on more than strictly scientific grounds, appealing to the supposedly adverse political and moral consequences of, say, promoting the idea that humans are nothing but evolved animals. In contrast, calls for the inclusion of creationism, while often agreeing with the spirit of the former proposal, grant that evolution has significantly increased our understanding of natural phenomena, but hold that the Neo-Darwinian explanatory framework may not be adequate, and in any case would benefit from regularly having to confront historically relevant alternatives. Most so-called creationist movements in today's world, including the campaign for ID, fit into this category.

The second confusion involves failing to distinguish between the actual quality of proposed course materials and their real or imagined ideological inspiration. A telling moment in *Kitzmiller* came when the defence forced Kenneth Miller to admit that his own best-selling biology textbook defined Darwin's theory of evolution in equivocal terms, similar to those found in heavily criticised creationist texts.[1] While Miller's concession mattered little to the outcome of the trial, it revealed that those peddling the scientific orthodoxy could get away with sloppier language than those who would contest it. Clearly, what is needed is a consistent standard for evaluating course content. Most regional and national educational authorities provide a proxy for this by bench-marking examination performance: however the course is delivered, students need to know X by school year Y. Proposed textbooks and curricula should continue to be judged along these lines, regardless of whether they are provided by evolutionists or creationists. Indeed, we need to take seriously the proposition that students can score well in an exam on their understanding of evolution after using a textbook of ID or even creationist provenance. Competence and belief are two separate matters, a point that should already be clear from the number of professorships held in comparative religion and even theology by people who are either agnostic or atheist. One can know evolutionary theory perfectly well – and recognise the reasons for its dominance in contemporary biological science – without personally believing that it is true, or at least the whole truth.

ID needs to revisit the intellectual schisms in biology that the Neo-Darwinian synthesis overcame in the middle third of the 20th century, versions of which still endure in the social sciences: qualitative vs. quantitative methods, field vs. lab research sites, macro vs. micro perspectives. To a large extent, the language of modern evolutionary theory papers over, rather than resolves, the divergent perspectives of these scientific cultures by portraying them as ultimately contributing to a common vision of reality that was first outlined in Darwin's *On the Origin of Species*. However, scientists continue to invest differently in this vision. Indeed, it is worth recalling the words of Ronald Fisher, the Christian eugenicist who first elaborated the workings of natural selection in terms of population-based statistical principles of hereditary transmission. As evidenced in this letter to Julian Huxley, written in 1930 (i.e. shortly before the consolidation of the Neo-Darwinian synthesis), Fisher was much more interested in humanity's capacity to partake of divine creativity in the lab than in following the divine path through history:

> If I had so large an aim as to write an important book on Evolution, I should have had to attempt an account of very much work about which I am not really qualified to give a useful opinion. As it is there is surprisingly little in the whole book [i.e. Fisher's own classic *Genetical Theory of Natural Selection*] that would not stand if the world had been created in 4004 BC, and my primary job is to try to give an account of what Natural Selection *must* be doing, even if it had never done anything of much account until now.[2]

Nowadays, molecular biologists who treat DNA as a 'clock' are committed to evolution only as a principle for the *ordering*, but not the *timing* – or, for that matter, the *producing* – of species. It makes little difference to them what historical circumstances are responsible for the sequence of mutations registered in DNA, or whether these mutations occurred over many million or only a few thousand years. Thus, questions surrounding the radiometric dating of fossils – that is, the use of the decay rates of various elements found in fossils to determine when organisms lived – matter little to these supposed 'evolutionists'. Indeed, the American Scientific Affiliation, the largest pro-Christian scientific association, has convened a study group on this matter, since questions of when and how evolution occurred are precisely what is under dispute between Neo-Darwinists and creationists, including ID theorists. Recognising these fault lines in the Neo-Darwinian synthesis, ID's goal should be more to divide than to displace. This is because much of the biology that currently flies under the banner of 'Darwinism' relies little, if at all, on the bone of contention between evolutionists and ID theorists and other creationists; namely, whether life has developed over a very extended timeframe through purely self-organising natural processes.

On the religious side, ID needs to reassert the specificity of the Abrahamic God as the implied intelligent designer. Without this specificity (which still allows for considerable theological dispute), the concept of an intelligent designer becomes devoid of content, adding to the suspicion that ID is no more than 'not-evolution'. In this spirit, ID's critics have proffered a 'flying spaghetti monster' and an 'orbiting teapot' as alternatives to a more biblically inspired deity. In response, ID defenders should openly confront the relatively recent anti-religious judicial reading of the US Constitution's separation of Church and state, which now excludes even religiously motivated views from public science education: the issue should not be whether ID is primarily science *or* religion, but whether it passes scientific muster *as* an openly religious viewpoint with scientific aspirations – a matter to be decided by actual educational practice.

Here theologians could do more to pull their weight. Vague, comforting talk about nature's wonder and divine consolation for human misery short-changes the specifically *cognitive* role that theology has played in the shaping of the modern scientific world view. Theology is nothing if not a literal 'science of God'. In particular, ideas about God as the intelligent designer are rooted in theodicy's concern with cosmic justice, the source of the secular ideas of optimisation and adaptation on which the likes of Richard Dawkins continue to trade, as noted in chapter 5. Indeed, given his disagreements with Gould and his enthusiasm for genetic enhancement, Dawkins arguably owes more to 18th-century secular theodicy than to Darwin's own 19th-century anti-theodicy.

Roaming from their sectarian comfort zones, theologians might also stress that the God of Abraham does not belong exclusively to those sections of Protestant Christianity keen on reforming secular America, but more generally presides over three great world religions: Judaism, Christianity and Islam. While a creative force is common to most cosmologies, it usually appears as either an overriding natural tendency or a supernatural entity beyond all human comprehension. In contrast, the God of Abraham distinguished humans from the rest of nature as a creature *in imago dei*. This provides a strong reason for believing that reality constitutes an intelligible universe, as opposed to one or more realities manufactured by one or more beings, with mind(s) so different from ours that trying to resolve the plethora of natural phenomena into a unified theory would constitute a waste of time and effort, given our ephemeral stay on Earth. The cost of this unified monotheism is, of course, the prospect of Unitarianism and its controversial humanism, as exemplified in such critical readers of the Bible as John Toland and Gotthold Lessing, who first popularised the classification of Judaism, Christianity and Islam as types of 'monotheism' during the 18th-century Enlightenment.[3]

But most of all, ID needs to reclaim the history of biology. Here, an elementary point in the sociology of knowledge comes in handy: if you want to reveal a group's ideological motivation, look at who needs to be ignored in order to enable them to claim originality for their ideas. You can be sure that those conspicuous by their absence were motivated differently. The point cuts both ways in the current evolution–ID debate. Both sides place considerable rhetorical stress on the alleged originality of Michael Behe, William Dembski and other fellows of Seattle's eclectically right-wing Discovery Institute. For evolutionists, this points to the marginal – for creationists, the radical – character of their ideas. Yet, either interpretation implausibly suggests that these Discovery Institute fellows would reinvent biology from scratch. ID theorists would be better served by writing themselves back into the history of mainstream science, where they have been responsible for concepts in taxonomy, morphology, physiology, genetics and biochemistry that are still very much taken for granted, *especially* at the level of the school biology textbook.

In this respect, the 'track record' of Neo-Darwinism is parasitic on prior creationist breakthroughs over which Neo-Darwinists now claim sole ownership, and which creationists have yet to claim back as their own. Moreover, the recovery of this history – ideally in textbook presentations of scientific reasoning – would demonstrate the power of creationist thought in what philosophers of science call the 'context of discovery', the source of scientific inspiration that should be the centre of gravity in science education. After all, theologically informed creationist premises have motivated the conduct of science, the results of which have been successfully used and built upon by both theists and non-theists. Especially in a time when pure science departments are closing for lack of enrolments, this is a potentially powerful selling point. Requiring students to leave their religion at the door of the science class is denying them a proven tool of systematic inquiry.

References

Introduction

1 For a sample of the recent debates concerning intelligent design, see William Dembski and Michael Ruse (eds.), *Debating Design* (Cambridge, UK: Cambridge University Press, 2004).

2 Thomas Kuhn, *The Structure of Scientific Revolutions* (second edn; original edn 1962) (Chicago, IL: University of Chicago Press, 1970). On this book's deep and largely pernicious influence on the public – and some scientists' – understanding of science, see Steve Fuller, *Thomas Kuhn: A Philosophical History for Our Times* (Chicago, IL: University of Chicago Press, 2000); *Kuhn vs. Popper: The Struggle for the Soul of Science* (Cambridge, UK: Icon, 2003).

Chapter 1

1 My sense of the elusiveness of consensus in science goes back to the only part of my doctoral dissertation that has made it into print, as chapter 9 of *Social Epistemology* (Bloomington, IN: Indiana University Press, 1988).

2 The idea that science constitutes a 'church' around which consensus must form was perhaps raised most vividly by Ernst Mach against Max Planck's claims for the orthodoxy of the atomic world view in physics in the early 20th century. See

Steve Fuller, *Thomas Kuhn: A Philosophical History for Our Times* (Chicago, IL: University of Chicago Press, 2000), chapter 2.

3 On the subtlety of Bacon's position, see John Henry, *Knowledge and Power: Francis Bacon and the Method of Science* (Cambridge, UK: Icon, 2002).

4 For an interesting attempt to analyse the idea of common mental preparation in terms of 'planning for spontaneity', which draws on Polanyi's association with the émigré Austrian economist Adolph Lowe in Manchester, see Mathew Forstater, 'Must spontaneous order be unintended? Exploring the possibilities for consciously enhancing creative discovery and imaginative problem-solving', in H.S. Jensen, L.M. Richter and M.T. Vendelø (eds.), *The Evolution of Scientific Knowledge* (Cheltenham, UK: Edward Elgar, 2003), pp. 189–208.

5 The political pressures faced by discipline-based science in Germany can be traced through the career of Max Planck. On Althoff's higher education policy, see Jürgen Backhaus (1993) 'The University as an Economic Institution: The Political Economy of the Althoff System', *Journal of Economic Studies* vol. 20 (4/5), pp. 8–29. On 'free inquiry' under Nazism, which capitalised on scientists' default tendency to focus simply on their own work, see Ute Deichmann, *Biologists under Hitler* (Cambridge, MA: Harvard University Press, 1999).

6 Steve Fuller, *Science* (Milton Keynes, UK: Open University Press, 1997), pp. 67–76.

7 Derek de Solla Price, *Little Science, Big Science* (London, UK: Penguin, 1963).

8 Steven Weinberg, *Dreams of a Final Theory*; (New York, NY: Atheneum, 1992).

9 Steve Fuller, *The Governance of Science* (Milton Keynes, UK: Open University Press, 2000), p. 136.

10 For two contrasting views of the 'hoax', see Andrew Ross (ed.), *Science Wars* (Durham, NC: Duke University Press, 1996), and Alan Sokal and Jean Bricmont, *Intellectual Impostures* (London, UK: Phaidon, 1998). See also Steve Fuller, *The Philosophy of*

Science and Technology Studies (London, UK: Routledge, 2006), pp. 102–114.

11 Characteristic of the considered STS response is Harry Collins and Robert Evans, 'The third wave of science studies: studies of expertise and experience', *Social Studies of Science*, 2002, vol. 32, pp. 235–96. The *locus classicus* for discussions of the 'academic left' is Paul Gross and Norman Levitt, *Higher Superstition* (Baltimore, MD: Johns Hopkins University Press, 1994).

12 This point is developed in Steve Fuller, *The Philosophy of Science and Technology Studies*, pp. 122–8.

13 Sheila Harkins, *Kitzmiller v. Dover Area School District, Transcript*, day 19, AM, 2 November 2005, p. 115.

14 John Dupré, *The Disorder of Things: Metaphysical Foundations of the Disunity of Science* (Cambridge, MA: Harvard University Press, 1993); Karin Knorr-Cetina, *Epistemic Cultures* (Cambridge, MA: Harvard University Press, 1999).

15 In the case of nano-biotechnology, I mean the 'converging technologies' agenda first advanced by the US National Science Foundation in 2002 and subsequently emulated by major science policy agencies around the world. It aims to focus bio-, info- and nano- technology research towards the 'enhancement of human performance'. For one of the original American statements on the subject, see Mihail Roco and William Sims Bainbridge, 'Converging technologies for improving human performance: Integrating from the nanoscale', *Journal of Nanoparticle Research*, 2002, vol. 4, pp. 281–95.

16 On the shifting politics of race, see Desmond King, *In the Name of Liberalism* (Oxford, UK: Oxford University Press, 1999); Nikolas Rose, *The Politics of Life Itself* (Princeton, NJ: Princeton University Press, 2007).

17 The finding and reactions to it were reported at the conference 'Beyond Belief: Science, Religion, Reason and Survival', held at the Salk Institute, La Jolla, CA, on 5–7 November 2006.

18 Sociologist Robert Merton called this tendency the 'principle of cumulative advantage'. For a critique, see Steve Fuller, *The Governance of Science*, chapter 5.

19 Robert May, 'Respect the facts', *Times Literary Supplement*, 6 April 2007. See also Ben Pile and Stuart Blackman, 'The Royal Society's "motto-morphosis"', *Spiked*, 15 May 2007. http://www.spiked-online.com/index.php?/site/earticle/3357/

20 Steve Fuller, 'The trouble with facts', *New Scientist*, 22 June 2002.

21 I discuss the issue in more detail in *Science vs. Religion? Intelligent Design and the Problem of Evolution* (Cambridge, UK: Polity Press, 2007), p. 131ff.

22 Apparently $1 million was less than half the billable fees, which allowed a moment of black humour in which the law firm supporting the ACLU declared its magnanimity for charging so *little*. See Christina Kauffman, 'Dover gets a million-dollar bill', *The York [Pennsylvania] Dispatch*, 22 February 2006. In September 2006, Indiana congressman John Hostettler managed to get a bill passed by the US House of Representatives to protect the 'public expression of religion'. Were the bill to become law, this form of economic intimidation would become illegal. However, the Senate has refused to vote on the bill, and Hostettler himself lost his seat in the 2006 midterm elections.

23 See for example Theodore Schick, 'Methodological Naturalism vs. Methodological Realism', *Philosophy*, vol. 3/2 (2000), pp. 30–37.

24 Interestingly, before they came to be seen as specifically 'scientific' philosophies, both naturalism and positivism were read as heretical religious positions, the one (associated with followers of Spinoza) identifying the deity with nature, the other (associated with followers of Comte) with a fully self-realised humanity. Those interested in a 20th-century expression of naturalism as a heretical religious doctrine should look to the works of George Santayana. In the case of positivism as a 20th-century religious heresy, readers are directed to the General

Semantics Movement, an intellectual forebear of contemporary Scientology.

25 Steven Weinberg and Lewis Wolpert, a physicist and a biologist respectively, are good examples of such distinguished high-profile scientists.

26 The concept of the 'meme', a (pseudo?) Darwinian attempt to explain the spread of ideas, has been most developed in this context – that is, one that treats the spread of religious ideas as a disease to be prevented, if not cured. See Daniel Dennett, *Breaking the Spell: Religion as a Natural Phenomenon* (New York, NY: Viking, 2006); Richard Dawkins, *The God Delusion* (New York, NY: Houghton Mifflin, 2006). Here a clever psychoanalyst might suspect the sublimation of Neo-Darwinism's eugenicist impulses at play in a world that prohibits their direct expression: rather than sterilise religious people, Neo-Darwinists would instead sterilise their minds.

27 Cornelia Dean, 'Believing Scripture but Playing by Science's Rules', *New York Times*, 12 February 2007. The case concerned a geologist, Marcus Ross, who trained at the University of Rhode Island and now teaches at Liberty University, Virginia, founded by the evangelist Jerry Falwell.

28 The great historian of the 'paranoid' and 'anti-intellectual' style of US politics is Richard Hofstadter.

Chapter 2

1 http://www.pamd.uscourts.gov/kitzmiller/decision.htm, on p. 64 of the judge's verdict. Thanks to Brian Thomasson, who drew attention to this quote in 'Arguing from the Evidence: The Correct Approach to Intelligent Design and the U.S. Courts', from the 'Darwinism after Darwin' conference, University of Leeds, 3 September 2007.

2 Stephen Jay Gould, *Rocks of Ages* (New York, NY: Vintage Books, 1999).

3 Jones revealed himself to be less than religiously devout in an interview published one week before his verdict. Laurie

Goodstein, 'Evolution Trial in Hands of Willing Judge', *New York Times*, 18 December 2005. Goodstein writes, 'Asked if he was religious, he said he attended a Lutheran church favoured by his wife, but not every Sunday'. After reading that, I realised that the Dover school board was doomed.

4 Robert Proctor, *The Nazi War on Cancer* (Princeton, NJ: Princeton University Press, 1999), pp. 126–33. Lorenz and Tinbergen shared the 1973 Nobel Prize in Physiology or Medicine.

5 Charles Darwin, *The Descent of Man and Selection in Relation to Sex* (second edn; original edn 1871) (London, UK: John Murray, 1882), especially pp. 65–96, 565, 576–7, 618–19.

6 On Darwin's views of vivisection, see Francis Darwin (ed.), *The Life and Letters of Charles Darwin, including an autobiographical chapter*, vol. 1 (London, UK: John Murray, 1887), pp. 201–10.

7 On science as itself posing a challenge to the natural ecology, see Harold Dorn, *The Geography of Science* (Baltimore, MD: Johns Hopkins University Press, 1991). One of the few people who recognises clearly the continuity between humanity's biblical sense of entitlement and the progressivism associated with the modern scientific world view – but only to condemn it – is the British political theorist, John Gray.

8 Francis Darwin (ed.), *The life and letters of Charles Darwin*, vol. 1, pp. 103–7.

9 David Livingstone, *Darwin's Forgotten Defenders* (Grand Rapids, MI: William Eerdmans, 1984), chapter 4.

10 Tertullian was giving a moralistic spin to Plato's fable of the Thracian slave girl's witty response to Thales' misadventure in *Theaetetus* 174a: she made it seem that Thales had merely put the cart before the horse in paying more attention to the starry heavens above than the Earth below. Interestingly, Plato's defenders have tended to read this passage as implicitly criticising the other-worldliness of *scientists*, not philosophers. In contrast, Tertullian criticises scientists like Thales for an all-too-worldly sense of curiosity.

11 See Barbara Benedict, *Curiosity: A Cultural History of Early Modern Inquiry* (Chicago, IL: University of Chicago Press, 2001).
12 On the vexed politics of science in medieval Islam and Christianity, see Toby Huff, *The Rise of Early Modern Science: Islam, China and the West* (Cambridge, UK: Cambridge University Press, 1993).
13 Readers familiar with the history and philosophy of science will recognise here my endorsement of the 'hypothetical modelling', or 'maker's knowledge' style of scientific thinking as discussed in Alistair Crombie, *Styles of scientific thinking in the European tradition: The history of argument and explanation especially in the mathematical and biomedical sciences and arts*, 3 vols. (London, UK: Duckworth, 1994).
14 Amos Funkenstein, *Theology and the Scientific Imagination* (Cambridge, UK: Cambridge University Press, 1986), pp. 26-32, 57–9. For the implications of the radical Scotist reading of Augustine for modern social theory (i.e. human dominion over nature), see John Milbank, *Theology and Social Theory* (Oxford, UK, Blackwell, 1990), pp. 302–6.
15 St Augustine, translated and annotated by J.H. Taylor, S.J., *The Literal Meaning of Genesis* (New York, NY: Paulist Press, 1982), Book 1, chapter 18, paragraph 37. The final sentence is omitted, e.g. in Francis Collins, *The Language of God: A scientist presents evidence for belief* (New York, NY: Free Press, 2006), p. 83.
16 On this specific point, see Margaret Schabas, 'Adam Smith's Debt to Nature', *History of Political Economy*, Annual Supplement to Volume 35 (2003), pp. 262–81. On the general connection between natural history and political economy, see the introduction to Wolf Lepenies, *Between Literature and Science: The Rise of Sociology* (Cambridge, UK: Cambridge University Press, 1988).
17 Staffan Müller-Wille, 'Nature as Marketplace', *History of Political Economy* Annual Supplement to Volume 35 (2003), pp. 154–72.

18 Lisbeth Koerner, *Linnaeus: Nature and Nation* (Cambridge, MA: Harvard University Press, 1999), p. 92. This book is generally recommended as the most knowing work on this greatest of applied theodicists.

19 Charles Darwin, *On the Origin of Species by Means of Natural Selection, or the preservation of favoured races in the struggle for life* (London, UK: John Murray, sixth edn, 1872: original edn 1859), p. 425.

20 Robert J. Richards, 'The Linguistic Creation of Man: Charles Darwin, August Schleicher, Ernst Haeckel, and the Missing Link in 19th-Century Evolutionary Theory', in M. Doerres (ed.), *Experimenting in Tongues: Studies in Science and Language* (Palo Alto, CA: Stanford University Press, 2002).

21 Schleicher's original use of the cladogram as a device for mapping linguistic migration, along with its racialist overtones, has been updated in recent years as the 'human genetic diversity project', spearheaded by the geneticist Luigi Cavalli-Sforza and embodied in the 'Genographic Project', to which anyone may contribute their family histories. See https://www3.nationalgeographic.com/genographic/overview.html

22 For an accessible introduction to Vico, see Isaiah Berlin, *Vico and Herder: Two Studies in the History of Ideas* (New York, NY: Viking, 1976).

23 David Hull, 'Darwin's science and Victorian philosophy of science' in J. Hodge and G. Radick (eds.), *The Cambridge Companion to Darwin* (Cambridge, UK: Cambridge University Press, 2003), pp. 168–91.

24 On the influence of inquisitorial legal systems on Bacon, see James Franklin, *The Science of Conjecture: Evidence and Probability before Pascal* (Baltimore, MD: Johns Hopkins University Press, 2001), pp. 217–18.

25 Whewell probably learned of Kant through his early mentor, the Romantic poet Samuel Taylor Coleridge. See Laura Snyder, *Reforming Philosophy* (Chicago, IL: University of Chicago Press, 2006), pp. 42–51.

26 Still the best introduction to this aspect of modern philosophy of science is Gerd Buchdahl, *Metaphysical Foundations of Natural Science* (Cambridge, MA: MIT Press, 1969).
27 Francis Bacon (ed. A. Johnston), *The Advancement of Learning*, (Oxford, UK: Oxford University Press, 1973), Book I.vi.16 (p. 42); Book II.vi.1 (p. 86). Cf. *Novum Organum*, Book I, section 89, in J. Spedding, R. Ellis, and D. Heath *(eds.), The Works of Francis Bacon*, 14 vols (London, UK: Longman and Co., 1857–74), vol. IV, p. 88.
28 Larry Laudan, *Science and Hypothesis* (Dordrecht, Netherlands: Kluwer, 1981), chapter 14.
29 John Passmore, *The Perfectibility of Man* (London, UK: Duckworth, 1970), chapter 10.
30 A typical popular account appears in Michael White, *Rivals: Conflict as the Fuel of Science* (London, UK: Secker & Warburg, 2001), pp. 57–96, in which Priestley's science is presented as having been compromised by his religious 'fanaticism'. Priestley's demotion was largely due to Kuhn's Harvard patron, James Bryant Conant, the university's president as well as a trained chemist, who devised the early history of science courses in which 'Priestley vs. Lavoisier' figured as a case study of how scientists' conceptual frameworks (aka paradigms) constrain how they see the world.
31 The double irony should not be lost that Islam, the culture that invented epistemic universalism, ended up being demonised as 'the other', and then resurrected as the poster-child of post-colonialism, via Edward Said's *Orientalism*, written by a Palestinian Christian, no less!
32 A good contrast between Grosseteste and Albertus over the concept of light is provided in Yael Raizman-Kezar, 'Plotinus's conception of unity and multiplicity as the root to the medieval distinction between *lux* and *lumen*', *Studies in History and Philosophy of Science* vol. 37 (2006), pp. 379–97.
33 On group–grid analysis, see Mary Douglas, *Purity and Danger* (London, UK: Routledge and Kegan Paul, 1966). Its most

creative application in the study of science has been made by David Bloor in virtually all his work, starting with *Knowledge and Social Imagery* (London, UK: Routledge and Kegan Paul, 1976).

34 Richard Rubinstein, *Aristotle's Children* (Orlando, FL: Harcourt, 2003), pp. 190–2.

35 Martyn Amos, *Genesis Machines: The New Science of Biocomputing* (London, UK: Atlantic Books, 2006).

36 Ray Kurzweil, *The Singularity is Near: When Humans Transcend Biology* (New York, NY: Viking, 2005).

37 Michael Walzer, *The Revolution of the Saints: A Study in the Origins of Radical Politics* (Cambridge, MA: Harvard University Press, 1965).

38 Henry Steele Commager, *The Empire of Reason: How Europe Imagined and America Realized the Enlightenment* (London, UK: Weidenfeld and Nicolson, 1978), chapter 2. Jefferson strongly encouraged Priestley to compose a demystifying commentary on the Bible to complement Jefferson's own rendition of Jesus' ascendancy in the Gospels – the so-called Jefferson Bible – as (very much like the American Revolution) an unintended consequence of a tax revolt. That Jefferson would even have written such a thing in the first place is indicative of the 'performative' sense of biblical literalism that is explored in chapter 7 of this book.

39 On Priestley's significance, see Friedrich Engels, *Ludwig Feuerbach and the End of Classical German Philosophy*, (Peking, China: Foreign Languages Press, 1976 original edn 1888), pp. 82–4, 143–5, 161–2. A linchpin in the trajectory from Priestley through Hegel to science as an instrument of human advancement is John Stallo, a German émigré to the US who was a leader of the 'Ohio Hegelians' in the late 19th century. His view was picked up and normalised for Anglophone audiences as 'operationalism' by Harvard's Percy Bridgman, whose 1946 Nobel Prize-winning work on the physics of high pressures included a failed, alchemically inspired attempt

to synthesise diamond. One similarly contrarian contemporary scientist who cites Bridgman as an inspiration is Martin Fleischmann. See his 'Background to Cold Fusion: Genesis of a Concept', *Proceedings of the Tenth International Conference on Cold Fusion* (Cambridge, MA: World Scientific, Inc., 2003). It is perhaps no coincidence that Bridgman supervised the PhD of Gerald Holton, a founder of the discipline of history of science in the United States. Since by the 1940s Bridgman was one of a dwindling number of successful scientists who continued to treat atoms as *merely* theoretical entities, he was clearly an instance of someone whose atavistic existence demonstrated how much things had changed. By living against the grain of history, Bridgman was a scientist of the sort that only a philosopher of science could love.

40 Roger Wood and Vitezslav Orel, 'Scientific breeding in Central Europe in the Early Nineteenth Century: Background to Mendel's Later Work', *Journal of the History of Biology*, vol. 38 (2005), pp. 239–72.

41 On the idea of a Linnaean research tradition, of which Mendel was a part, see Robert Olby, 'Mendel, Mendelism and Genetics' (1997), at http://www.mendelweb.org/MWolby.html, accessed 9 May 2007.

42 Michael Ruse, 'Evolutionary Biology and the Question of Trust', in N. Koertge (ed.), *Scientific Values and Civic Virtues* (Oxford, UK: Oxford University Press, 2005), pp. 99–119.

43 Margaret Mann Lesley, 'Mendel's Letters to Carl Nägeli', *The American Naturalist*, vol. 61, no. 675 (July–August 1927), pp. 370–78.

44 Michael Ruse, *Mystery of Mysteries: Is Evolution a Social Construction?* (Cambridge, MA: Harvard University Press, 1999), pp. 83–8.

Chapter 3

1 Peter Dear, *The Intelligibility of Nature* (Chicago, IL: University of Chicago Press, 2006).

2 It would be more accurate to say that the Church is the grand master of what Jean-Paul Sartre called the morality of 'dirty hands'. It has always pursued a diverse ideological portfolio, allowing the promotion of any viewpoint as long as it might increase its marginal advantage – and certainly not disadvantage it – in its worldwide mission. For example, Copernican astronomy was promoted by Jesuit missionaries in China at the same time that Galileo was being persecuted for declaring its truth at home. In the former case, the idea was to destabilise the Chinese emperor, whereas in the latter it was to prevent the Pope's own destabilisation. A comparable logic applies to the US prosecution of the so-called 'war on terror', at least under the influence of neo-conservative ideology: as civil liberties are curtailed for Americans under the rubric of 'homeland security', the people of Iraq are lectured on the freedoms that will flow naturally from adhering to market economies and the rule of law.

3 The great theorist of the scapegoat in sacred and secular literature is René Girard.

4 See Steve Fuller, *Kuhn vs. Popper: The Struggle for the Soul of Science* (Cambridge, UK: Icon, 2003), chapters 10–11.

5 On Weaver generally, see David F. Noble, *The Religion of Technology: The Divinity of Man and the Spirit of Invention* (Harmondsworth, UK: Penguin, 1997), pp. 177–8. On Weaver's role at the Rockefeller Foundation, see Michel Morange, *A History of Molecular Biology*, (Cambridge, MA: Harvard University Press, 1998), chapter 8; Robert Kohler, *Partners in Science: Foundations and Natural Scientists, 1900–1945* (Chicago: University of Chicago Press, 1991), chapter 10; Daniel Kevles, 'Foundations, Universities, and Trends in Support for the Physical and Biological Sciences, 1900–1992', *Daedalus* 121/4 (1992), p. 206ff.

6 Jennifer Hecht, *The End of the Soul: Modernity, Atheism and Anthropology in France* (New York, NY: Columbia University Press, 2003). In this context, French intellectuals who remained

friendly to theology took solace in the work of Henri Bergson, which asserted the superiority of Lamarck to Darwin as promising a convergence of the divine and human intellects. The ultimate beneficiary of this pro-science, pro-theology tradition was Pierre Teilhard de Chardin, author of *The Phenomenon of Man* (New York, NY: Harper & Row, 1955).

6 S.J. Gould, 'Impeaching a self-appointed judge' (a review of Phillip Johnson's *Darwin on Trial*), in *Scientific American*, vol. 267 (1992), pp. 118–21.

7 From Theodosius Dobzhansky, 'Nothing in biology makes sense except in light of evolution', *The American Biology Teacher*, March 1973 (3), pp. 125–9.

9 Theodosius Dobzhansky, *The Biology of Ultimate Concern* (New York, NY: New American Library, 1967), pp. 108–37.

10 John Hedley Brooke, *Science and Religion: Some Historical Perspectives* (Cambridge, UK: Cambridge University Press, 1990), pp. 283–4.

11 A latter-day intellectual descendant of Mivart is Simon Conway Morris, Professor of Evolutionary Palaeobiology at Cambridge University, perhaps Stephen Jay Gould's leading opponent on his own turf. Where Gould claimed that replaying the history of life on Earth would probably result in a radically different evolutionary trajectory, Morris argues that at most the result would be a change in the 'tempo and mode' of natural history, to recall George Gaylord Simpson's phrase, but not in the overall direction of evolution's trajectory. See Simon Conway Morris, *Life's Solution: Inevitable Humans in a Lonely Universe* (Cambridge, UK: Cambridge University Press, 2003).

12 On the long-term significance of the Condorcet–Malthus dispute over the feasibility/desirability of indefinite human expansion, see Steve Fuller, *The New Sociological Imagination* (London, UK: Sage, 2006), chapter 13.

13 E.O. Wilson, *The Creation: An Appeal to Save Life on Earth* (New York, NY: W.W. Norton, 2006).

14 See Freeman Dyson, 'Our Biotech Future', *The New York Review of Books*, 19 July 2007, which draws heavily on the work of the microbiologist Carl Woese.

15 Francis Collins, *The Language of God: A scientist presents evidence for belief* (New York, NY: Free Press, 2006).

16 Paul Slovic, 'Genocide: When Compassion Fails', *New Scientist*, 7 April 2007.

17 See Peter Singer, *A Darwinian Left* (London, UK: Weidenfeld and Nicolson, 1999); Steven Pinker, *The Blank Slate: The Modern Denial of Human Nature.* (New York, NY: Vintage, 2002). Watson's offending remarks appeared in the *Sunday Times* (London, UK), 14 October 2007. One sociologist who has systematically theorised – and condemned – this tendency is Norman Geras, in *The Contract of Mutual Indifference* (London, UK: Verso, 1998).

18 Richard Dawkins, *The Selfish Gene* (Oxford, UK: Oxford University Press, 1976), pp. 101–3.

19 To his credit, Dawkins does not shirk the potential consequences of this perspective vis-à-vis his own life. See Richard Dawkins, 'The word made flesh', in the *Guardian* (London), 27 December 2001.

20 See Fuller, *The New Sociological Imagination*. In metaphysical terms, the shift from liberalism to meta-liberalism amounts to converting a 'one-many' to a 'part-whole' relationship. Thus, a universalism of the human (at the expense of the non-human) life is replaced by a holism encompassing all forms of life (at the expense of suboptimal individuals, including humans).

21 Collins, *The Language of God*, p. 149. Collins inveighs equally against sociobiologists and cultural relativists, whom he sees as challenging theistic evolution from the naturalistic and the humanistic sides. His ultimate bugbear would be Edward Westermarck, the first social anthropology chair at the London School of Economics, who embodied *both* sides over half a century before either became widespread.

22 Ibid., p. 199.

23 For example Jerry Fodor, who has long been a scourge of evolutionary explanations, especially as purveyed by fellow philosopher and Dawkins cheerleader Daniel Dennett. He regularly writes on these matters for the *Times Literary Supplement* and the *London Review of Books*. Fodor is probably the follower of Noam Chomsky who, over the years, has remained most faithful to Chomsky's own Cartesian 'methodological solipsist' approach to the nature of mind. Steven Pinker, by contrast, is probably the best-known Chomskyan who has accommodated his position to Neo-Darwinism.

24 Francisco Ayala, 'Darwin's Greatest Discovery: Design without a Designer', *Proceedings of the National Academy of Sciences*, vol. 104 (15 May 2007), pp. 8567–73.

25 See especially Michel Foucault, *The Order of Things* (New York, NY: Random House, 1970; original edn 1966).

26 The missing link between evolutionary naturalism and postmodernism is, of course, Nietzsche. This theme is pursued in my *New Sociological Imagination*.

27 Eric Schliesser, 'Hume's Newtonianism and Anti-Newtonianism', at http://plato.stanford.edu/entries/hume-newton/

28 Larry Laudan, *Science and Hypothesis* (Dordrecht, Netherlands: Kluwer, 1981), chapter 6.

29 http://anthro.palomar.edu/synthetic/synth_2.htm. The content of the first paragraph, which presents an historical and then a conceptual account of the Hardy–Weinberg principle, reverses the original order for purposes of continuity.

30 In this definition, 'allele' is the technical name for 'gene', referring specifically to what the DNA sequence on a chromosome codes for. Thus, for some genes, there are 'dominant' and 'recessive' alleles that the organism may possess, a set of which constitutes its 'genotype'.

31 For an astute comparison of biology and economics along these lines, made with a view to debunking the ontological pretensions of 'equilibrium' in both, see Alexander Rosenberg, *Economics: Mathematical Politics or Science of Diminishing*

Returns? (Chicago, IL: University of Chicago Press, 1993), pp. 193–4. For a more respectful comparison along much the same lines, see Jack Hirschleifer, 'Economics from a Biological Standpoint', *The Journal of Law and Economics*, vol. 20/1 (1977), pp. 1–52.

32 See especially Nancy Cartwright, *How the Laws of Physics Lie* (Oxford, UK: Oxford University Press, 1983) and *The Dappled World* (Cambridge, UK: Cambridge University Press, 1999).

33 Robert Rosen, *Essays on Life Itself* (New York, NY: Columbia University Press, 1999), chapter 19, 'Bionics revisited'.

34 The research received popular coverage in Will Knight, 'Beetle's jet may inspire new engines', *New Scientist*, 9 December 2003.

35 E.O. Wilson, *Consilience: The Unity of Knowledge* (New York, NY: Alfred Knopf, 1998).

36 The economic implications of biomimetics have been explored under the rubric of 'natural capitalism'. A popular account is provided in J.M. Benyus, *Biomimicry: Innovation Inspired by Nature* (New York, NY: HarperCollins, 1997).

37 William Paley, *Natural Theology* (1800), chapter 2, paragraph 3.

38 Stephen Meyer, 'DNA and the Origin of Life: Information, Specification and Explanation', in J.A. Campbell and S. Meyer, (eds.), *Darwinism, Design and Public Education* (Lansing, MI: Michigan State University Press, 2003), pp. 223–85.

39 Charles Darwin, *On the Origin of Species by Means of Natural Selection, or the preservation of favoured races in the struggle for life* (London: John Murray, 1859), p. 81.

40 Ronald Fisher, *The Genetical Theory of Natural Selection* (New York, NY: Dover, 1930). The Christian subtext of Fisher's statistical presentation is grasped in William Dembski, 'The logical underpinnings of intelligent design' in W. Dembski and M. Ruse (eds.), *Debating Design* (Cambridge, UK: Cambridge University Press, 2004), chapter 17, but ignored in criticism, e.g. Elliott Sober, 'The Design Argument', Ibid., chapter 6.

41 Charles Darwin, *On the Origin of Species by Means of Natural Selection, or the preservation of favoured races in the struggle for life*

(London, UK: John Murray, sixth edn 1872; original edn 1859), p. 49.

42 On various anticipations of Nazism by German philosophical and scientific Darwinism, see Richard Weikart, *From Darwin to Hitler* (New York, NY: Macmillan, 2005).

Chapter 4

1 Jeffrey Herf, *Reactionary Modernism* (Cambridge, UK: Cambridge University Press, 1984).

2 David Livingstone, *Darwin's Forgotten Defenders* (Grand Rapids, MI: Eerdsman, 1984), chapter 5.

3 James D. Watson, *DNA: The Secret of Life* (New York, NY: Alfred Knopf, 2003), pp. 238–41.

4 Sean Carroll, *The Making of the Fittest: DNA and the Ultimate Forensic Record of Evolution* (New York, NY: Norton & Norton, 2006).

5 Henry Gee, *In Search of Deep Time: Cladistics, the Revolution in Evolution* (London, UK: HarperCollins, 2000).

6 Contrast Philip Kitcher, *Abusing Science: The Case against Creationism*, (Cambridge, MA: MIT Press, 1982); Philip Kitcher, *Living with Darwin: Evolution, Design, and the Future of Faith* (Oxford, UK: Oxford University Press, 2007).

7 P.J. Bowler, *The Non-Darwinian Revolution* (Baltimore, MD: Johns Hopkins University Press, 1992).

8 This idea as a counter-Kuhnian philosophy of science is developed in Steve Fuller, *Thomas Kuhn: A Philosophical History for Our Times* (Chicago, IL: Chicago University Press, 2000), chapter 8.

9 Stephen Meyer, Scott Minnich, Jonathan Moneymaker, Paul Nelson and Ralph Seelke, *Explore Evolution: The Arguments for and against Neo-Darwinism* (Melbourne, Australia: Hill House, 2007).

10 Karl Popper, *The Poverty of Historicism* (New York, NY: Harper & Row, 1957).

11 Richard Hofstadter, *Social Darwinism in American Thought* (Philadelphia: University of Pennsylvania Press, 1944). Before Hofstadter, 'Social Darwinism' was a neutral expression for the application of Darwin-inspired ideas to social life.
12 Will Provine, 'Progress in Evolution and Meaning in Life', in M. Nitecki (ed.), *Evolutionary Progress*. (Chicago, IL: University of Chicago Press, 1998), pp. 49–72.
13 A critically sympathetic account of Popper's various changes of mind concerning Darwin's theory of evolution is provided in Costas Krimbas, 'In Defence of Neo-Darwinism: Popper's "Darwinism as a Metaphysical Program" Revisited', in R. Singh et al., (eds.), *Thinking about Evolution: Historical, Philosophical, and Political Perspectives* (Cambridge, UK: Cambridge University Press, 2001), pp. 292–308.
14 Michael Behe, *Darwin's Black Box* (New York, NY: Simon & Schuster, 1996), p. 39, quoting Darwin, *On the Origin of Species*, p. 189.
15 Mark Ridley, *How to Read Darwin* (London, UK: Granta, 2005), p. 61.
16 Steve Jones, *Almost Like a Whale: The Origin of Species Updated* (London, UK: Doubleday, 1999), p. 20.
17 The golden age of social science methodology, which took seriously the epistemological problems of operationalising complex concepts and integrating findings from multiple methods, occurred at the Berkeley psychology department in the 1940s, then associated with E.C. Tolman and Egon Brunswik. The collected writings of the main student of that period are still worth reading, not least because of their broad endorsement of a 'Darwinian' perspective: Donald T. Campbell, *Methodology and Epistemology for Social Science* (Chicago, IL: University of Chicago Press, 1988).
18 Kenneth Miller, *Finding Darwin's God* (New York, NY: HarperCollins, 1999), p. 143.
19 Ibid., p. 144.

20 Michael Behe, *The Edge of Evolution* (New York, NY: Simon & Schuster, 2007).

21 For a secular version of these traditional theologically inspired concerns, see Francis Fukuyama, *Our Posthuman Future* (London, UK: Picador, 2003).

22 Ernst Cassirer, *The Problem of Knowledge: Philosophy, Science and History since Hegel* (New Haven, CT: Yale University Press, 1950), pp. 188–216.

23 Miller, *Finding Darwin's God*, p. 145.

24 For an extended critical analysis of their debate, see Kim Sterelny, *Dawkins vs. Gould: Survival of the Fittest* (Cambridge, UK: Icon, 2001).

25 For an elaboration of this distinction, see Steve Fuller, *Science vs. Religion? Intelligent Design and the Problem of Evolution* (Cambridge, UK: Polity Press, 2007), pp. 35–43.

26 Quoted in Ian Hacking, *The Emergence of Probability* (Cambridge, UK: Cambridge University Press, 1975), p. 172. Newton wrote this to his protégé Richard Bentley, who was selected in 1692 to give the first Robert Boyle Lecture, an annual series – continued to this day at St Mary le Bow Anglican Church in London – designed to show how the latest science upholds the Christian faith against unbelievers.

27 Michael Behe, *Kitzmiller v. Dover Area School District, Transcript*, day 11, 18 October 2005, pp. 38–42.

28 Recently, a medieval historian has incisively critiqued the historical pretensions of evolutionary psychology, while urging his colleagues to take both Neo-Darwinism and neuroscience more seriously: Daniel Lord Smail, *On Deep History and the Brain* (Berkeley, CA: University of California Press, 2008).

29 Michael Ruse, *Darwin and Design: Does Evolution Have a Purpose?* (Cambridge, MA: Harvard University Press, 2003), p. 302. McCosh's views are interestingly contrasted with A.D. White's. It is also worth noting that in the unpublished 1844 essay that is normally seen as the prototype for *Origin of Species*, Darwin quite clearly tried to cast natural selection as

a principle comparable to Newtonian gravity – i.e. something that acts at a distance, as a supernatural force. He didn't like the idea that individual species or aspects of them might be selected for through the individual choice of the Creator, but he did like the idea that there might be some scientific way of saying that all the variation exhibited in nature is the result of the same general law operating under specific circumstances. One could say that Darwin was a believer in 'holistic design' (i.e. at the macro level of the species, an ecosystem or the Earth itself, as opposed to the more micro level of the individual, gene etc.).

30 Douglas Walton, 'Historical Origins of Argumentum ad Consequentiam', *Argumentation*, vol. 13 (1999), pp. 251–64.

31 The most accessible recent source on the full range of astrologers' activities in their scientifically most respectable heyday is Anthony Grafton, *Cardano's Cosmos: The Worlds and Works of a Renaissance Astrologer* (Cambridge, MA: Harvard University Press, 2000).

32 John Maynard Keynes, 'Newton, the Man'. Lecture written for delivery to the Royal Society of London on the occasion of the 300th anniversary of Sir Isaac Newton's birth (1942). Due to the Second World War and Keynes' death shortly afterwards, the lecture was delivered by his brother in 1946. See http://www-groups.dcs.st-and.ac.uk/~history/Extras/Keynes_Newton.html

33 David Hull, *The Metaphysics of Evolution* (Albany, NY: SUNY Press, 1989), pp. 162–78; E.O. Wilson, 'The meaning of biodiversity and the tree of life', in Joel Cracraft and Michael Donoghue (eds.), *Assembling the Tree of Life* (Oxford, UK: Oxford University Press, 2004), pp. 539–42. On how this might affect the placement of humans in the tree of life, see Rebecca Cann, Mark Stoneking, Allan Wilson, 'Mitochondrial DNA and Human Evolution', *Nature*, vol. 325 (1987), pp. 31–6.

Chapter 5

1 The strategy of reducing design to pattern is characteristic of William Dembski's more mathematically adept critics, such as Wesley Elsberry and Jeffrey Shallit. Yet intelligent design lurks even here, since 'pattern' is etymologically akin to 'patron', meaning 'he who advances the cause'.

2 For a brief but suggestive discussion of what its author somewhat dismissively calls 'the obsessional search for meaning', see Jon Elster, *Sour Grapes: Studies in the Subversion of Rationality* (Cambridge, UK: Cambridge University Press, 1983), p. 102–3. Elster coins the term 'biodicy' to describe Darwin's replacement of theodicy. (Richard Dawkins is today's leading biodicist.) Elster distinguishes Malebranche's and Leibniz's versions of theodicy: Malebranche saw theodicy as being about the nature of divine creativity (i.e. the nature of the trade-offs God made in solving the ultimate optimisation problem), whereas Leibniz saw it as being about humanity's moral instruction (i.e. that we cannot know the good unless the bad also exists as a contrast). But Leibniz's view can be extended to spur us to do better and hence perfect creation *à la* bionics and biomimetics – i.e. God provides exemplars of how things go *wrong*, as well as right.

3 One of the very few clear recognitions of the connections between divine optimisation, global equilibrium, biomimetic engineering and contemporary ID appears in Bernadette Bensaude-Vincent, 'Reconfiguring Nature through Syntheses: From Plastics to Biomimetics' in B. Bensaude-Vincent and W. Newman (eds.), *The Artificial and the Natural: An Evolving Polarity* (Cambridge, MA: MIT Press, 2007), pp. 293–312. Perhaps unsurprisingly, Bensaude-Vincent's expertise lies in the history of chemistry rather than biology.

4 I. Bernard Cohen, *Science and the Founding Fathers* (Cambridge, MA: Harvard University Press, 1995).

5 See Milič Čapek, *The Philosophical Impact of Contemporary Physics* (Princeton, NJ: Van Nostrand, 1961), on Henri Bergson and Alfred North Whitehead in their attempts to deal with Einstein and Darwin as posing related problems to Newton's mechanical world view.

6 Arthur Silverstein, 'Darwinism and Immunology: From Metchnikoff to Burnet', *Nature Immunology*, vol. 4 (2003), pp. 3–6.

7 Robert Proctor, *Racial Hygiene: Medicine under the Nazis* (Cambridge, MA: Harvard University Press, 1988). The contemporary implications of this point are developed in Steve Fuller, *The New Sociological Imagination* (London, UK: Sage, 2006), chapter 14.

8 Herbert Simon, *The Sciences of the Artificial* (Cambridge, MA: MIT Press, 1977), pp. 220–9).

9 See Herbert Vetter (ed.), *Notable American Unitarians* (Cambridge, MA: Harvard Square Press, 2007).

10 Simon, *The Sciences of the Artificial*, pp. 200–2.

11 Pat Langley, Herbert Simon, Gary Bradshaw and Jan Zytkow, *Scientific Discovery: Computational Explorations of the Creative Processes* (Cambridge, MA: MIT Press, 1987).

12 See especially James Secord, *Victorian Sensation: The Extraordinary Publication, Reception, and Secret Authorship of 'Vestiges of the Natural History of Creation'* (Chicago, IL: University of Chicago Press, 2001).

13 Sean Carroll, *Endless Forms Most Beautiful: The New Science of Evo-Devo* (New York, NY: Norton, 2005), pp. 71–2. The image of evolution as 'tinkerer' (or '*bricoleur*') was popularised by François Jacob, in 'Evolution and tinkering', *Science*, vol. 196 (1977), pp. 1161–6. Jacob drew it from Claude Levi-Strauss' account of the 'savage mind', which allows the heretical esoteric reading that we have been created to 'civilise' God's savage handiwork.

14 Thomas Goudge, 'Evolutionism', in Philip Wiener, (ed.), *Dictionary of History of Ideas* (New York, NY: Charles Scribners

& Sons, 1973), vol. 2, pp. 174–189. I discuss the implications of this point in *The New Sociological Imagination* (London, UK: Sage, 2006), part III.

15 See Huxley's 1893 Romanes Lecture, 'Evolution and Ethics', http://aleph0.clarku.edu/huxley/CE9/E-E.html

16 Jerome Schneewind, *The Invention of Autonomy* (Cambridge, UK: Cambridge University Press, 1997), pp. 264–71.

17 Charles Darwin, *On the Origin of Species by Means of Natural Selection, or the preservation of favoured races in the struggle for life* (London, UK: John Murray, sixth edn, 1872: original edn 1859), pp. 188–9.

18 George Basalla, *The Evolution of Technology* (Cambridge, UK: Cambridge University Press, 1988).

19 David Knight, *Science and Spirituality: The Volatile Connection* (London, UK: Routledge, 2004), pp. 49–51.

20 The *locus classicus* for the shift in intuitions between the 19th and the 20th centuries about the need for models in scientific reasoning is Mary Hesse, *Models and Analogies in Science* (South Bend, IN: University of Notre Dame Press, 1963).

21 If the reader doubts that computers should be seen as natural successors to the 19th century's fixation on mechanical models for representing subtle physical processes, see Jon Agar, 'What difference did computers make?', *Social Studies of Science*, vol. 36 (2006), pp. 868–907. Agar shows that historically computers were aids more to imagination than to computation, since typically the relevant computations had already been performed by more strictly human means before being transferred to computers.

22 For a state-of-the-art account of computer simulations as high-tech exercises in 19th-century model-building (*à la* Faraday and Maxwell) that under ideal circumstances would be carried out as laboratory experiments, see Nancy Nersessian, *Creating Scientific Concepts* (Cambridge, MA: MIT Press, 2008). The implied sequence, expressed in Latin to give it gravitas, is as follows: *in vivo* is virtualised *in vitro*, which in turn is

virtualised *in silico*. On computer simulation as the emerging ultimate target of all scientific research, see John Horgan, *The End of Science: Facing the Limits of Science in the Twilight of the Scientific Age* (Reading, MA: Addison-Wesley, 1996).

23 R. Lenski, C. Ofria, R. Pennock, C. Adami, 'The Evolutionary Origin of Complex Features', *Nature*, vol. 423 (2003), pp. 139–44. See also Robert Pennock, *Kitzmiller v. Dover Area School District, Transcript*, day 3, 28 September 2005, pp. 91–2.

24 Michael Behe, 'Kenneth R. Miller and the Problem of Evil', Michael Behe's Amazon Blog, 24–26 October 2007. See http://www.amazon.com/gp/blog/A3DGRQ0IO7KYQ2

25 For the US National Science Foundation's attempt to realise Weaver's ID-oriented vision through the promotion of nanotechnology, see Mihail Roco and William Sims Bainbridge, 'Converging technologies for improving human performance: Integrating from the nanoscale', *Journal of Nanoparticle Research*, 2002, vol. 4, pp. 281–95.

Chapter 6

1 E. Meléndez-Hevia, T. Waddell and M. Cascante, 'The puzzle of the Krebs citric acid cycle: assembling the pieces of chemically feasible reactions, and opportunism in the design of metabolic pathways during evolution', *Journal of Molecular Evolution*, vol. 43 (1996), p. 202. Emphasis as in the original.

2 Kenneth Miller, *Finding Darwin's God* (New York, NY: HarperCollins, 1999), p. 151.

3 Darwin uses the word 'cell' only to refer to a unit of honeycomb produced by bees. And while Darwin makes a couple of incidental references to microscopes, he never uses the adjective 'microscopic' to refer to a reality that lies beneath ordinary observation. These facts are among the many that can be discovered by searching Cambridge University's online archive of Darwin's published works and letters: http://darwin-online.org.uk/

4 Perhaps the most philosophically developed version of this position appears in Charles Hartshorne, *The Divine Relativity: A Social Conception of God* (New Haven, CT: Yale University Press, 1948).

5 Lily Kay, *Who Wrote the Book of Life: A History of the Genetic Code* (Palo Alto, CA: Stanford University Press, 2000), p. 95.

6 Norbert Wiener, *The Human Use of Human Beings* (Boston, MA: Houghton Mifflin, 1950), p. 51. For the philosophical and political context of this conceptualisation, see Philip Mirowski, *Machine Dreams: Economics Becomes a Cyborg Science* (Cambridge, UK: Cambridge University Press, 2002), especially p. 55. For the most complete expression of Wiener's theological interests in cybernetics, see his *God and Golem, Inc.*, (Cambridge, MA: MIT Press, 1964), which won a US National Book Award.

7 Warren Weaver, 'Science and the Citizen', *Science*, vol. 126 (3285), pp. 1225–9 (13 December 1957). Weaver's piece was published only a few weeks after the USSR had launched Sputnik, the first artificial space satellite, which is generally seen as having definitively turned the Cold War into a science war.

8 Simcha Jong, 'Traditional University Organization and the Emergence of New Technological Fields: Cambridge University and the Rise of Biotechnology', paper delivered to a European Union-sponsored workshop, 'Converging Science and Technologies: Research Trajectories and Institutional Settings', Vienna, 15 May 2007.

9 Nicolas Rasmussen, 'The Mid-century Biophysics Bubble: Hiroshima and the Biological Revolution in America, Revisited', *History of Science*, vol. 35 (1997), pp. 245–91.

10 This distinction has been traced to a fundamental difference in orientation to the study of life, already present in the 17th century, between 'preformationists' and 'epigeneticists', whose bone of contention was the trajectory of the individual organism, not an entire species, let alone the entire history of life. The

preformationists, like today's population geneticists, treated genes as templates for specific traits, whereas the epigeneticists, like today's molecular biologists, treated genes as potential vehicles for the expression of various traits depending on the context. See Lenny Moss, *What Genes Can't Do* (Cambridge, MA: MIT Press, 2003).

11 Contrast Richard Lewontin, *Biology as Ideology: The Doctrine of DNA* (Cambridge, MA: Harvard University Press, 1991) with Walter Gilbert, 'Towards a Paradigm Shift in Biology', *Nature*, vol. 349, 10 January 1991, p. 99.

12 Michel Morange, *A History of Molecular Biology* (Cambridge, MA: Harvard University Press, 1998), p. 249.

13 Robert Rosen, *Essays on Life Itself* (New York, NY: Columbia University Press, 1999), chapter 1, 'The Schrödinger Question'.

14 For an imaginative yet rigorous account of what a full-blown version of the revolution in molecular biology might look like, such that life might be artificially constructed without genes, see Adrian Woolfson, *Life without Genes: The History and Future of Genomes*. (London, UK: HarperCollins, 2000).

15 Erwin Schrödinger, *What is Life? With Mind and Matter and Autobiographical Sketches* (Cambridge, UK: Cambridge University Press, 1967), pp. 86–90.

Chapter 7

1 See Steve Fuller, *Social Epistemology* (Bloomington, IN: Indiana University Press, 1988), chapter 5, where I observe the role that the biblical translation scholar, Eugene Nida, played in the formulation of the two most influential theories of interpretation in analytic philosophy, Quine's indeterminacy thesis and Kuhn's incommensurability thesis.

2 Stephen Stigler, 'John Craig and the probability of history: from the death of Christ to the birth of Laplace', *Journal of the American Statistical Association*, vol. 81 (1986), pp. 879–87. For a similar perspective, see the introduction to my *Philosophy of*

Science and Its Discontents, (New York, NY: Guilford Press, second edn 1993; original edn 1989).

3 Peter Dear, '*Totius in verba*: Rhetoric and Authority in the Early Royal Society', *Isis*, vol. 76 (1985), pp. 145–61.

4 Perhaps the most profound antirealist philosopher of recent times has developed this position, taking literally the idea that God creates via *logos* and hence as a literary creator. See Michael Dummett, *Thought and Reality* (Oxford, UK: Oxford University Press, 2006), pp. 96–110.

5 Major discussions of mimesis include W.K.C. Guthrie, *A History of Greek Philosophy* (Cambridge, UK: Cambridge University Press, 1962), vol. 1, pp. 230–1; Erich Auerbach, *Mimesis: The Representation of Reality in Western Literature* (Princeton, NJ: Princeton University Press, 2003, original edn 1946); Paul Ricoeur, *The Rule of Metaphor* (Toronto, ON: University of Toronto Press, 1977) and Robert Rosen, *Essays on Life Itself* (New York, NY: Columbia University Press, 1999), chapter 7, 'On Psychomimesis'.

6 The most thorough development of this thesis is Peter Harrison, *The Bible, Protestantism and the Rise of Natural Science* (Cambridge, UK: Cambridge University Press, 1998).

7 On the conventions surrounding revelation and concealment during the Scientific Revolution, see Steven Shapin, *A Social History of Truth* (Chicago, IL: University of Chicago Press, 1994).

8 Jerome Schneewind, 'The Divine Corporation and the History of Ethics', in R. Rorty, J. Schneewind and Q. Skinner (eds.), *Philosophy in History* (Cambridge, UK: Cambridge University Press, 1984), pp. 173–92.

9 Richard Dawkins, 'Foreword' to J. Burley, (ed.), *The Genetic Revolution and Human Rights* (Oxford, UK: Oxford University Press, 1999), p. vi.

Conclusion

1 Kenneth Miller, *Kitzmiller v. Dover Area School District, Transcript*, day 2, AM, 27 September 2005, pp. 5–7.
2 Fisher to Huxley, 6 May 1930. Quoted in Michael Ruse, *Monad to Man* (Cambridge, MA: Harvard University Press, 1996), p. 295.
3 An excellent account of the ramifications of this move is Jonathan Israel, *Radical Enlightenment* (Oxford, UK: Oxford University Press, 2001).

Index

The Intellectual

Steve Fuller

'Smart and sassy.'
Times Higher Education Supplement

'I devoured this in one sitting. It's packed with juicy nuggets of genuine intellectual nourishment on every page.'
Dylan Evans

Christopher Hitchens, Richard Dawkins, Germaine Greer, Martin Amis ... with their regular TV appearances, newspaper columns and soundbites in times of crisis, intellectuals are essential characters in the drama of modern life. What makes them tick?

With Niccolò Machiavelli's notorious tract on statecraft, *The Prince*, firmly in mind, Steve Fuller dissect what it means to be an Intellectual – a Prince for our time.

What is an intellectual? What distinguishes them from philosophers, scientists, politicians and entrepreneurs? How do they stalk their quarry? What codes do they live by? Why are they happy to be insulted as long as they are not ignored?

Get to the heart of what it means to be an intellectual and meet exemplars along the way like Voltaire, Sartre, Chomsky, Fukuyama and many more. Think you would recognise an intellectual if you saw one? Entertain hopes of intellectualism yourself? Here's the essential guide.

Kuhn vs. Popper

The Struggle for the Soul of Science

Steve Fuller

'The very book I've been needing to read for ages!'
Jenny Uglow

'An eloquently written book, offering new and interesting perspectives on the moral and social ramifications of this debate.'
New Scientist

In 1965 Thomas Kuhn and Karl Popper met at the University of London to stage what became the most momentous philosophical debate of the century. At heart was the soul of science itself.

Popper pinned the future of science on scientists having the freedom to test their theories to the point of being false. But this required an 'open society' that tolerates error, even in established authorities.

Kuhn, in contrast, reflected the heads-down Cold War mentality that scientists should not question authority in their own fields or in society at large – unless absolutely necessary. Those rare occasions count as proper 'scientific revolutions'.

Kuhn, painted as the young radical against Popper as the conservative, won the battle. Steve Fuller argues forcefully, however, that these caricatures of Kuhn and Popper's positions are fundamentally flawed – and that the wrong man won.

The first popular account of this landmark confrontation, *Kuhn vs. Popper* retells the story of the clash, its background, and its legacy to our understanding of science.

Paperback: UK £7.99, Canada $17.00
ISBN 978-1840467-22-2

Atom

Piers Bizony

'Insightful and compelling ... a human drama wrought with frustration, love, guilt and genius.'
New Scientist

'Bizony elevates [his subject] to high drama, and shows how we are still being confronted with bizarre discoveries and questions in this brand of science.'
Good Book Guide

A BBC TV tie-in, *Atom* is the thrilling human story of atomic theory – one of modern science's most exciting, fundamental and mind-bending ideas.

No one ever expected the atom to be as bizarre, as capricious and as weird as it turned out to be. Its tale is one riddled with jealousy, rivalry, missed opportunities and moments of genius.

Piers Bizony tells the story of the young misfit New Zealander, Ernest Rutherford, who showed that the atom consisted mainly of empty space, a discovery that turned 200 years of classical physics on its head, and the brilliant Dane, Niels Bohr, who made the next great leap into the incredible world of quantum theory.

Yet he and a handful of other revolutionary young scientists weren't prepared for the shocks that Nature had up her sleeve. At the dawn of the Atomic Age, a dangerous new force was unleashed with terrifying speed ...

Paperback: UK £8.99, Canada $18.00
ISBN 978-1840468-73-1

Proust and the Squid

Maryanne Wolf

'An entertaining, comprehensive, delightfully clear account of how our brain allowed us to become word magicians. A splendid achievement!'
Alberto Manguel, author of *A History of Reading*

'We were never born to read', says Maryanne Wolf. 'No specific genes ever dictated reading's development. Human beings invented reading only a few thousand years ago. And with this invention, we changed the very organisation of our brain, which in turn expanded the ways we were able to think, which altered the intellectual evolution of our species.'

In *Proust and the Squid*, Maryanne Wolf explores our brains' near-miraculous ability to arrange and re-arrange themselves in response to external circumstances. She examines how this 'open architecture', the elasticity of our brains, helps and hinders humans in their attempts to learn to read, and to process the written language. She also investigates what happens to people whose brains make it difficult to acquire these skills, such as those with dyslexia.

Wolf, a world expert on the reading brain, brings both a personal passion and deft style to this, the story of the reading brain.

Hardback: UK £12.99
ISBN 978-1840468-67-0

The History of Britain Revealed

M.J. Harper

'Unusual, funny and provocative.'
New Statesman

'Mind-blowing, incredibly entertaining stuff.'
Daily Mail

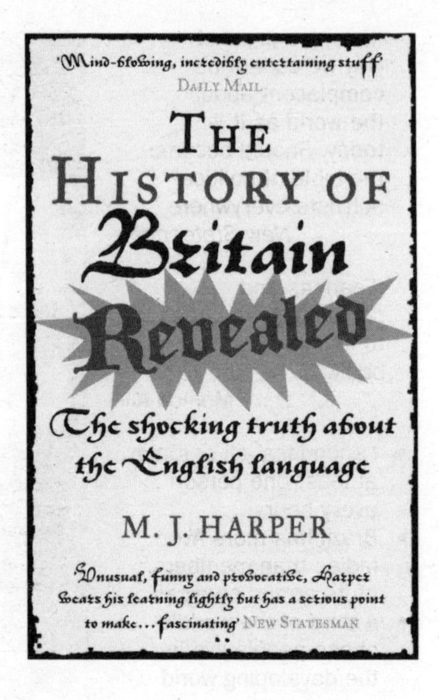

Do you think you know where the English language came from? Think again.

In gloriously corrosive prose, M.J. Harper destroys the cherished national myths of the English, the Scots, the Welsh, the Irish and – to demonstrate his lack of national bias – the French. In doing so he also shows that most of the entries in the Oxford English Dictionary are wrong, the whole of British place-name theory is misconceived, Latin is not what it seems, the Anglo-Saxons played no major part in our history or language, and Middle English is a wholly imaginary language created by well-meaning but deluded academics.

Iconoclasic, unsentimental and truly original, *The History of Britain Revealed* will change the way you think about history, language and much else besides. It is an essential but rarely comforting read for anyone who believes that history matters.

Paperback: UK £7.99, Canada $16.00
ISBN 978-1840468-35-9

The New Rome

The Fall of an Empire and the Fate of America

Cullen Murphy

'Provocative and lively'
New York Times

'A nuanced and convincing view'
Esquire (US)

'Mesmerising'
Financial Times

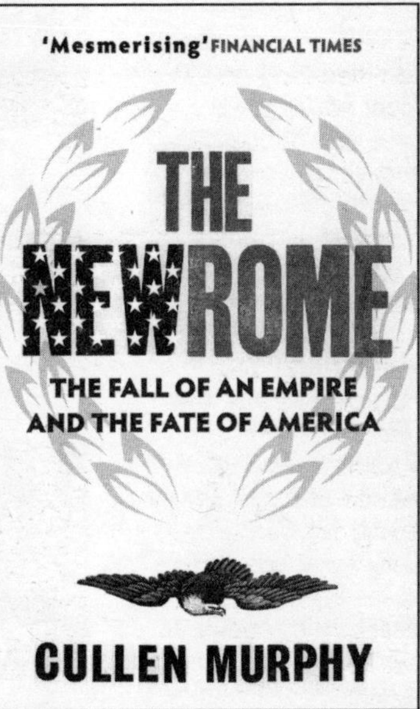

The rise and fall of ancient Rome has always been a metaphor for America, but here Cullen Murphy ventures past the obvious, bringing the brutal colours of Rome and the complexities of today's USA together in this beautifully written, intelligent and hugely readable book.

He explores how the two populations saw their political elites, and the insular cultures of Washington and Rome. He looks at the consequences of military overstretch and the widening gap between military and civilian society. Murphy sees both states weakened through 'privatisation' and vexed by the paradoxical issue of borders. Pressingly, he argues that America most resembles Rome in the burgeoning corruption of its government and in its arrogant ignorance of the world outside; in these conditions, idealism, however well-meant, can too easily be a form of blindness.

Lively and richly peppered with historical stories, Murphy's book brings the ancient world to life, and casts today's biggest superpower in a provocative new light.

Hardback: UK £14.99
ISBN 978-1840468-87-8

Can You Trust the Media?

Adrian Monck with Mike Hanley

The media dominates our lives. We give more time to viewing, surfing, listening and reading than we do to our families and friends. It's a relationship supposedly built on trust – and it's a relationship currently in crisis.

TV's fake phone-ins, phoney footage from royal reality shows, reporters resorting to phone-bugging to get stories – is there anything left in the media we can believe?

As audiences wonder which way to turn, former TV news boss and award-winning journalist Adrian Monck turns an insider's eye on the scandals that have sucked the public's trust from the media.

Underneath it all he argues that as we dither about trust, the media doesn't really care. Editors and proprietors want your time, attention and money ... and if the truth gets stretched in the process, then so be it.

But in the interactive Internet world, is there anything we can do about this? Online readers are increasingly shaping the media they consume. But will this act as a bulwark against the lies and liberties; or even spur those on top to pay attention to the public debate? *Can You Trust The Media?* looks at the forces that have shaped the news, and those that are remaking it.